T

THE CURRENT INTERPRETATION
OF WAVE MECHANICS:
A CRITICAL STUDY

Original title
ÉTUDE CRITIQUE DES BASES DE L'INTERPRÉTATION
ACTUELLE DE LA MÉCHANIQUE ONDULATOIRE
GAUTHIER-VILLARS, PARIS, 1963

Translated by
EXPRESS TRANSLATION SERVICE
28 ALEXANDRA ROAD
WIMBLEDON, ENGLAND

THE CURRENT INTERPRETATION OF WAVE MECHANICS
A CRITICAL STUDY

by

LOUIS DE BROGLIE

Member of the Académie Française
Permanent Secretary of the Académie des Sciences
Professor at the Sorbonne, Paris (France)

with a chapter by

J. ANDRADE E SILVA

ELSEVIER PUBLISHING COMPANY
AMSTERDAM — LONDON — NEW YORK
1964

ELSEVIER PUBLISHING COMPANY
335 JAN VAN GALENSTRAAT, P.O. BOX 211, AMSTERDAM

AMERICAN ELSEVIER PUBLISHING COMPANY, INC.
52 VANDERBILT AVENUE, NEW YORK, N.Y. 10017

ELSEVIER PUBLISHING COMPANY LIMITED
12B, RIPPLESIDE COMMERCIAL ESTATE
RIPPLE ROAD, BARKING, ESSEX

LIBRARY OF CONGRESS CATALOG CARD NUMBER 64-18508

WITH 8 ILLUSTRATIONS

ALL RIGHTS RESERVED
THIS BOOK OR ANY PART THEREOF MAY NOT BE REPRODUCED IN ANY FORM,
INCLUDING PHOTOSTATIC OR MICROFILM FORM,
WITHOUT WRITTEN PERMISSION FROM THE PUBLISHERS

Contents

PREFACE vii

CHAPTER I. ON THE NATURE OF CORPUSCULAR AND WAVE PHENOMENA

1. Observation of the microphysical world 1
2. Statistical nature of wave phenomena. Examination of the concept of complementarity 6
3. Is the wave of wave mechanics objective of subjective? 8

CHAPTER II. THE FORMALISM AND THE USUAL INTERPRETATION OF WAVE MECHANICS

1. The wave equations of wave mechanics 11
2. Localisation principle and the principle of spectral resolution 13
3. Can the two principles stated above be unified? 17
4. Uncertainty relationships 18
5. The usual interpretation of the formalism of wave mechanics 20

CHAPTER III. DIFFICULTIES RAISED IN THE PRESENT THEORY BY THE HYPOTHESIS THAT THE PARTICLE IS NOT CONSTANTLY LOCALISED IN SPACE

1. Einstein's objection 24
2. Schrödinger's objection 26
3. Another form of the preceding objections 28
4. Renninger's negative-result experiment 29
5. An attempt to remove the foregoing difficulties 31
6. Einstein's argument 34

CHAPTER IV. SUMMARY EXPLANATION OF THE THEORY OF THE DOUBLE SOLUTION

1. Basic concepts 37
2. Introduction of a random element into the foregoing concepts 41
3. Einstein's views on waves and particles. Introduction of non-linearity into the Theory of the Double Solution 43
4. The form of the u-wave and the relationship between the u-wave and the Ψ-wave 46
5. A new way of considering the guidance formula 48

CHAPTER V. CRITICAL STUDY OF CERTAIN POINTS OF THE USUAL INTERPRETATION OF WAVE MECHANICS

1. Statement of the Theory of Transformations 50
2. Critique of the Theory of Transformations 52
3. Fundamental importance of the q-representation. Actual and predicted probabilities 55
4. Significance of the uncertainty relationships 57
5. The statistical structure of wave mechanics. 59
6. Impossibility of obtaining interference fringes and of determining, at the same time, the trajectory of the particle 61
7. Concerning an article by Max Born 68

CHAPTER VI. WAVE MECHANICS OF SYSTEMS OF PARTICLES AND THE THEORY OF THE DOUBLE SOLUTION

1. Statement of the problem. 71
2. Present state of the problem. Andrade e Silva's Thesis 73
3. On an early paper by Darwin. 77

CHAPTER VIII. REMARKS ON SYSTEMS OF IDENTICAL PARTICLES
(by J. Andrade e Silva)

1. General considerations 81
2. Introduction of the concept of transitory states. 84
3. Demonstration of the symmetry of wave functions for bosons 88

APPENDIX. NEGATIVE PROBABILITIES AND THE THEORY OF THE DOUBLE SOLUTION. 92

BIBLIOGRAPHY 95

Preface

For more than thirty years the majority of theoretical physicists have joined forces in an interpretation of quantum physics and wave mechanics which originates from concepts due to Niels Bohr and his followers (the Copenhagen School). This interpretation appears to adapt itself perfectly to the elegant and precise formalism which is currently used in quantum mechanics and leads to predictions which are generally in excellent agreement with experiment.

After my initial work on wave mechanics, I was led to a view of the wave–particle duality of matter which was completely different from that of the Copenhagen School. However, I was soon forced to abandon this view by the great difficulties which it presented, and finally accepted the interpretation which is now orthodox and which I have since taught.

On returning to my original ideas during the last ten years, I have arrived at the conviction that the usual formalism, though strict in appearance and leading generally to precise conclusions, does not provide a profound and a truly convincing explanation of the physical reality on the sub-microscopic scale.

Among the questions which this sudden change has led me to pose here is naturally this one: if the present orthodox ideas are inadequate, then why should eminent thinkers come to accept them, and why have I been resigned to them for so many years? The obvious answer is that once the formalism of quantum mechanics is accepted (and it *is* satisfactory in so many respects), it appears to lead naturally to these orthodox ideas. Having taught the currently accepted interpretation for a considerable time, I was in the position to undertake a thorough criticism and thereby bring to light the inadequacies and obscurities in the accepted reasoning, which appeared to be so decisive. In particular, particle localisation, which in the final analysis is the only phenomenon which we are able to observe indirectly at the microphysical level, appeared to me to be particularly unsatisfactory in that it had succeeded in making the concept of the localised particle disappear from the theory. In the text that follows I have given some examples of the paradoxes which result from this state of affairs.

On reflection, I have occasionally felt that some of the existing arguments in favour of the current interpretation can in fact be regarded as more favourable towards my own ideas; many examples of this will be found in the present monograph.

In this booklet, I have reviewed the interpretation of wave mechanics which I proposed earlier under the name of the Theory of the Double Solution, but have also taken into consideration the more recent evolution of my thoughts on this subject. In particular, I have stressed the importance of introducing the random element into the theory, which corresponds to the hypothesis of the "subquantised medium" of Bohm and Vigier. On this hypothesis, even an isolated particle must be considered as being in permanent contact with a "hidden thermostat", the seat of which would be what we call the "vacuum". I have in fact more and more realized that the original form which I gave to the Theory of the Double Solution should be completed by the addition of this random element. The dynamics of a particle, which I described by the "guidance formula", must thus be augmented by the addition of the thermodynamics of the particle, for which I am beginning to visualise a large field of enquiry. Thus, I draw near to an idea of Einstein's who with his remarkable physical intuition saw the intervention of something akin to Brownian motion in the behaviour of the particle in quantum physics. One is thus led to the consideration of fluctuations of a kind studied in statistical thermodynamics.

Nevertheless, though I have felt obliged to review in the following pages the principles of the new interpretation of the wave–particle dualism, this is not the principal aim of this work. Its main intention is to try to convey to the reader my growing conviction that the arguments on which the currently accepted interpretation of physics is based are not as decisive as they appear to be, but on the contrary, contain many significant loopholes. I have the growing impression that the majority of physicists who have yielded to these exaggerated abstract tendencies have too easily desisted from forming an intelligible representation of the phenomena of quantum physics.

In order to complete the text of this booklet Mr. João Luis Andrade e Silva has kindly written a complementary chapter entitled *Comments on Systems of Identical Particles*. I should like to thank him sincerely for his valuable collaboration,

1st July, 1962
LOUIS DE BROGLIE

CHAPTER I

On the Nature of Corpuscular and Wave Phenomena

1 – *Observation of the Microphysical World*

The microphysical world, that is to say, the world of atoms and fundamental particles, escapes our direct observation. How can we understand it? It seems that we should reply to this question by saying: we understand it only through the intermediary of "observable particle localisations", i.e. phenomena in which a particle acting at the microphysical level initiates an observable effect by a chain reaction. In order to define this more precisely, let us consider the following example. Suppose that a photon arrives in the emulsion of a photographic plate and gives rise to the photoelectric effect in one of the emulsion atoms, which results in the emission of a fast electron. The electron proceeds to ionise the neighbouring atoms and this promotes the emission of other electrons which in their turn ionise other atoms. This "snowball" process initiates chemical reactions in the photographic emulsion which have the result that, after development, a black spot of *macroscopic dimensions* appears on the plate and can be observed directly. If the particle instead of being a photon, as in the foregoing example, is an electron or another microparticle, it is always a process similar to the above chain reaction which enables its presence to be detected, e.g. in a plate, on a screen, and so on. This process of "observable particle localisation" seems to me to be the *only* process which enables us to understand what takes place at the microphysical level. The chain reaction initiated by the original particle is a kind of amplification process which indicates the occurrence of an event, which in itself is inaccessible, to direct observation.

The track left by a particle in a Wilson cloud chamber constitutes an example of observable particle localisations being produced in succession along a trajectory. In the supersaturated gas in the Wilson chamber, the incident electron undergoes successive interactions with the gas atoms, giving rise to a chain process which ultimately results in the formation of small water droplets which form the particle track. An intuitive interpretation of the observed phenomenon is immediately apparent. A similar explanation applies to events which takes place in a photographic emulsion, or in a bubble chamber, in which the passage of a particle is directly revealed.

Having acknowledged this fundamental concept, it will be appropriate to add some important comments.

The first point is that the result of the chain reaction (black spot on a photographic plate, etc.) is ultimately observed on the scale of a fraction of a millimetre, but the individual event which initiates the chain reaction takes place on the microphysical scale, on which typical linear dimensions are at the most of the order of the atomic radius (10^{-8} to 10^{-9} cm) but can be of the order of the nuclear radius (10^{-12} to 10^{-13} cm). We thus come to the conclusion that when a micro-particle initiates a phenomenon which is observable by a chain reaction, it always brings about an effect at a very short distance—a distance of microphysical order—on another particle, or on an atomic system. Whether we represent this effect by a potential function, or by the quite artificial procedure of virtual photon exchange, we must always attribute to it a range, a radius of action, which is very small ($\leq 10^{-8}$ cm). I shall add, without completely developing this point, that the effect must be considered as extremely localised even in the case of a photon interacting with a particle *. The idea that observable particle localisation—the only window which opens into the microphysical world for us—involves a very close *proximity* between the particle initiating the observable phenomenon and some other unit of the microphysical world, seems to me to be essential. All acceptable theories must feature it, and we shall see an example of it at the end of this volume when we shall consider an interesting paper of Darwin's on the representation of collision phenomena in configuration space.

The second important point is concerned with the dynamic variables of a particle, especially its energy and momentum. Contrary to what has sometimes been written, it seems to me that the energy or the momentum of a particle should never be measured by allowing it to undergo

---

* In modern theories of the interaction between electromagnetic radiation and charged particles, the interaction term always contains, under a more or less dissembled form, a factor of the type $\delta(\mathbf{R}-\mathbf{r})$, where $\mathbf{R}$ and $\mathbf{r}$ define the positions of the photon and the corpuscle. This amounts to admitting implicitly that the photon only acts on the particle if its position coincides almost exactly with that of the particle.

an interaction with a macroscopic body and then studying the recoil of the body: this recoil, by reason of its smallness, would in fact escape all observation. Actually, in order to measure the energy or the momentum of a particle, it is always necessary to utilise the interaction between this particle and another particle of atomic magnitude, and to observe the recoil by means of one or more observable localisations of the type previously described. The correlation established after the collision between the two microphysical units by means of the conservation of energy and momentum, will provide us with information about the dynamic variables of the particle under consideration.

As an example, we can take the Compton effect in a Wilson cloud chamber. An X-ray photon of known frequency ν_0 (the energy $h\nu_0$ and the momentum $h\nu_0/c$ are also known) reaches an electron at rest at a point O. After interaction between the photon and the electron, the

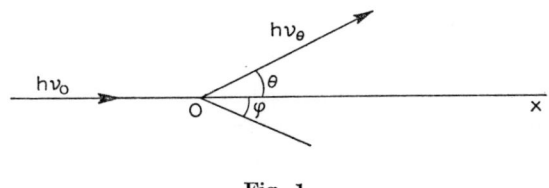

Fig. 1

photon moves away from the point O with an energy $h\nu_\theta$ and a momentum $h\nu_\theta/c$ in a direction making an angle θ with the incident direction Ox. Neither θ nor ν_θ are observable because the X-ray photon is too penetrating to produce observable effects in the gas within the chamber. The recoiling electron undergoes collisions with the gas atoms, which initiate condensations of water droplets and trace out the trajectory of the recoil electron. This trajectory has a zig-zag shape because of collisions between the electron and the gas atoms, but the tangent at the origin of the trajectory gives the angle φ between the initial electron recoil velocity and the Ox-axis.

The application of the laws of conservation of energy and momentum to the event leads to the following two well known relationships:

$$\nu_\theta = \frac{\nu_0}{1+\alpha(1+\cos\theta)} = \frac{\nu_0}{1+2\alpha\sin\dfrac{\theta}{2}}, \tag{1}$$

$$1-\cos\theta = \frac{2}{1+(1+\alpha)^2\tan^2\varphi} \quad \text{or} \quad \cot\varphi = (1+\alpha)\tan\frac{\theta}{2}, \tag{2}$$

where $\alpha = h\nu_0/mc^2$ and m is the mass of the electron. Since the angle φ can be observed, equation (2) can be used to calculate the value of θ. Substitution of this value into equation (1) enables us to calculate ν_θ. Thus, the observable localisations of the electron in the Wilson cloud chamber finally enable us to calculate the angle θ, which is unobservable, and also the unknown quantities $h\nu_\theta$ and $h\nu_\theta/c$, i.e. the energy and momentum of the photon after the event.

We may conclude from the foregoing discussion that every physical process which enables us to measure the dynamic variables of a particle, always results in an observable particle localisation of the type described above (or a succession of localisations of this type), subject to the final conservation of energy and momentum. Any process of measurement of a dynamic variable, such as the energy and momentum of a particle, is a complex and indirect process which necessarily utilises direct observation of particle localisations. This will enable us, when we make our critical study of the "Theory of Transformations", to throw doubt on the absolute equivalence which it postulates between the p and the q representations.

We again emphasise an important point. At the instant when an observable particle localisation is initiated, the particle should be found in very close proximity (at a distance of less than 10^{-8} cm) from the microphysical system upon which it acts. Moreover, what we call the "dimensions" of a micro-particle are certainly less than 10^{-12} cm. This corresponds to a very strict localisation of the particle in the wave train with which it is associated. Experiments show that the wave trains corresponding to photons of light are several million wavelengths long (i.e. several metres long), which is very different from 10^{-12} or even 10^{-8} cm. For electrons, recent experiments (Möllensted in Germany, Faget and Fert in Toulouse) have shown that the wave trains also have a length equal to very many times the wavelength (which is here of the order of 10^{-9} cm) and hence the wave train is very much longer than 10^{-12} or even 10^{-8} cm. It is this extreme localisation of the particle in its wave train, clearly revealed by the observable localisation, which constitutes, as we shall see, one of the principal difficulties in interpreting the wave-particle dualism.

Another important point is that the localisation of a particle is frequently defined in terms of its passage through a hole pierced in a screen. The smaller the aperture, it is said, the more precise is the localisation of the particle in the plane of the screen. Actually, there is no true localisation, since nothing can be observed as the particle passes through the aperture. Moreover, if the wave train is long, it is not known at what instant the particle has passed through the hole. The passage of the particle through the hole amounts to a restriction on the propagation of the wave associated with the particle, and is not a true particle localisation. The passage of the particle through the aperture may be contrasted with the photoelectric effect which is produced by a photon acting on an atom, or the ionisation of an atom by an electron, which are true observable particle localisations.

Another approach, which is often stated and which enters, as we shall see, into the Theory of Transformations, consists in comparing particle localisation with a contraction of the associated wave train to extremely small dimensions, represented approximately by a Dirac function $\delta(\mathbf{R}-\mathbf{R}_0)$. But such a contraction cannot be achieved experimentally and offers nothing but the abstract image of a wave train reduced to infinitely small dimensions.

Finally, I should like to emphasise again that the importance of the measuring apparatus is often exaggerated in the analysis of observations of microphysical entities. Very often there is *no* measuring apparatus in the true sense of the word. Where is the measuring apparatus when the arrival of a photon or an electron produces a local blackening on a photographic plate which can be examined visually, or when a physicist observes an electron track formed by the droplets in a Wilson cloud chamber? Certainly, in these two cases, and in other similar cases, there is always an experimental *device*: a photographic plate is placed in the way of the photon, an electron is admitted to the Wilson chamber, and so on. But, strictly speaking, there is no measuring apparatus. However, in certain cases a measuring apparatus can be inserted: for example, the blackening of the photographic plate may be assessed by means of a densitometer, or the photoelectric effect may be demonstrated by allowing the photoelectrons to operate an electron multiplier, which produces a current of sufficient intensity to be measurable with a galvanometer, and so on. But in all these cases, the measuring apparatus does not take part until the end of the observable localisation process, when the chain reaction will have reached a stage at which it can be measured by an ordinary measuring apparatus.

I believe that the analysis of particle localisation phenomena given above is important and merits a thorough examination. Such analyses should always be made without the introduction of any abstract mathematical formalism which would tend to mask the significance of the facts and thus adversely affect their interpretation.

2 – *Statistical Nature of Wave Phenomena. Examination of the Concept of Complementarity*

So far, we have only considered the concepts of the particle and the observable particle localisation. The concept of an associated wave is forced upon us by the phenomena of interference and diffraction. In order to account for the nature of these phenomena, we recall first of all that when photons or electrons fall on a photographic plate, they produce an effect resulting from the observable localisation processes which we described in the previous paragraph. This can be demonstrated experimentally. For example, it has been known for a considerable time that every light photon falling on a photographic emulsion gives rise to a photoelectric effect which initiates a local chemical reaction. This reaction eventually results in the reduction of silver bromide and in the formation of a small black spot on the negative (Silberstein, Vavilov).

Let us now consider a photographic plate on which the appearance of interference fringes or a diffraction pattern has resulted from the arrival of photons or electrons. Although the arrival of each particle produces only a small black point on the plate, the overall effect of the successive arrival of a large number of particles is the appearance of an interference or a diffraction pattern. Wave phenomena are thus essentially statistical in nature, since their appearance results from the arrival on the plate of a large number of particles which are distributed in proportion to the intensity of the wave.

This statistical interpretation of wave phenomena leads to a probabilistic picture of the wave. In fact, it has been possible to obtain interference phenomena by using ordinary light of very low intensity with very long exposure times. The intensity was in fact so low that not more than one photon at a time arrived at the plate. These experiments were carried out by Taylor in 1909 and repeated by Dempster and Batho in

1927. The same result has been obtained more recently (1949) in the U.S.S.R. by Suchkin and Fabrikant, who investigated electron diffraction patterns, and can be considered as verified for all particles. The only interpretation which can be given to this result is as follows: Each particle and the associated wave train falling on the diffraction target undergo diffraction as predicted by the wave theory. After a long period of time, when a large number of particles have arrived in succession at the plate, the observed particle localisations are finally distributed so that their number is proportional to the wave intensity. One is thus led to a wave mechanics in which a wave is associated with every particle in such a way that the intensity of the wave is a measure of the probability that the particle will produce an observable localisation at a point in space. We shall presently examine the difficulties which are raised by this interpretation of the probability wave, but for the moment let us consider the concept of complementarity by Niels Bohr.

According to Bohr, the wave and the particle are "complementary aspects" of physical reality. They appear in turn on different experiments in such a way that when one of them is manifested, the other is absent altogether. Thus, the two irreconcilable complementary aspects never enter into conflict with each other. The idea of complementarity has been very successful, and attempts have even been made to extrapolate it in a most dangerous manner out of the realm of physics into biology, sociology, psychology, and so on. I have, for a long time, adopted the idea of complementarity in the realm of quantum physics, whilst at the same time realising that it was inadequate. In recent years, I have been led to regard the concept of complementarity with increasing suspicion.

Many authors have described the concept of complementarity by saying that the fundamental units of light and of matter are proteiform, so that they appear to us sometimes in the form of waves and sometimes in the form of particles. This statement seems to me to be completely inaccurate. Thus, consider a photographic plate exhibiting interference fringes: the wave aspect of particles is clearly manifested, but the particle aspect is also present, since we know that the fringes have been produced on the plate by a succession of individual particle localisations. In other words, there exists on the plate a set of fringes which represents the wave aspect, but each black fringe is formed by an assembly of small black points which represent the particle aspect. Thus, the particle aspect and the wave aspect are presented together on the same plate, but the first is due to *individual* effects while the second is due to a *statistical* effect. We are thus faced with the absence of the single entity which allegedly assumes the particle and the wave aspects in turn.

We shall return later (in Chapter V) to the impossibility of the simultaneous observation of interference phenomena and the localisation of the particle on its trajectory, but for the present we must conclude that the concept of complementary can only be maintained if we are content to assign to it the following meaning: in order to understand completely the properties of particles, the localised-particle and extended-wave representations must be employed simultaneously. In this form the principle of complementarity is acceptable, but it is quite trivial and contributes no explanation whatsoever to the mystery of the unity of waves and particles. It seems to me to be a long way from having the deep philosophical significance which many authors attribute to it.

We now come to one of the most important questions, and one of the strangest riddles, posed by the interpretation of wave mechanics.

3 – *Is the Wave of Wave Mechanics Objective or Subjective?*

In the foregoing discussion, the intensity of the wave associated with a particle, i.e. the square of its amplitude, was assumed to be a measure of the probability of an observable localisation of the particle. This seems to indicate that a subjective character must be attributed to the wave. In my opinion, a probability representation implies partial ignorance and must therefore be subjective in character. I have said *partial* ignorance because total ignorance would not enable us to construct any probability representation whatsoever; in order to construct such a representation, it is essential to possess some information of an objective nature.

A very simple example will illustrate this point. Suppose a table has two drawers, one on the right and the other on the left. We know that one of the drawers contains a billiard-ball. This is already objective information, but it does not enable us to construct a probability law for the presence of the billiard-ball in one drawer or the other. But let us suppose that we are informed that the billiard-ball has been put into one of the drawers in accordance with the following procedure. The person who put the ball in the drawer first threw it on to a rotating roulette wheel which had an equal number of red and black compartments. He then decided to put the ball into the right-hand drawer if it stopped on a red compartment and into the left-hand drawer if it stopped on a black compartment. That is all we know, but this is objective information and it enables us to construct a probability distribution for the presence of the ball in the two drawers. The probability is in fact $\frac{1}{2}$ for each of the two drawers. This probability also represents the state of our ignorance about the position of the billiard-ball, and is thus essentially subjective in nature. If we open the right-hand drawer and find the billiard-ball, the probability distribution is sharply modified: it becomes 1 for the right-hand drawer and 0 for the left-hand drawer. This sharp modification shows clearly that except in the case of certainty, a probability distribution is the representation of our partial ignorance.

Thus, it seems that the probability representation must be regarded as subjective. But this point of view, which at first appears to be obtrusive, also raises considerable difficulties. It is, in fact, quite difficult to deny an entirely objective character to a wave which propagates in space in accordance with definite laws, which is reflected at mirrors, and which is diffracted when it encounters obstacles. How could this wave impart to the particle a privileged position in interference and diffraction phenomena if it exists only in our imagination? Moreover, the energies of stationary states depend on the conditions of propagation of the wave in any domain of space and on the corresponding boundary conditions. Is it reasonable that the form of the stationary wave corresponding to a proper frequency (eigenvalue) should impose on the particle a quantized value of its energy if this wave is purely subjective?

We now came to the heart of the problem: as the representation of a probability, the wave must be regarded as subjective, but insofar as it determines physical phenomena, it must also be objective. Authors who deal with this problem perpetually oscillate between the two contradictory viewpoints, feeling that neither can be rejected.

In 1926–1927, I had perceived a solution to this riddle which has, in recent years, appeared to me again to be the only reasonable one. This solution involves the existence of two different waves, one of which is objective and represents a physical reality which, because it is intimately linked with the particle, enables its behaviour to be determined. The other is a subjective construction, which is based on the information which we possess about the objective wave and provides us with a probability representation for the particle. I gave the name "Theory of the Double Solution" to this interpretation of wave mechanics because it led me to envisage two different solutions which are intimately linked with the wave equation.

We shall return later to the principles of the Theory of the Double Solution, but first we must summarise the accepted interpretation of wave mechanics in greater detail and thus lay the foundations of our subsequent critique.

CHAPTER II

The Formalism and the Usual Interpretation of Wave Mechanics

1 – *The Wave Equations of Wave Mechanics*

I shall not recount in detail here the origin of wave mechanics, but shall merely recall that according to my own early work and that of Schrödinger, a propagation equation can be used as the basis of wave mechanics. The form of this equation is suggested by the relationship between the motion of the particle and the propagation of the "monochromatic plane wave" ($W = h\nu$, $p = h/\lambda$), which was established in my early papers (Notes of 1923, Thesis of 1924), and also by the fact that in the limit of the geometrical optics approximation, wave mechanics must be consistent with the Hamilton-Jacobi theory of classical analytical mechanics.

The wave equation or equations used today depend on the species of the particle under consideration. For particles with zero spin, for which relativity corrections may be considered to be negligible, a single wave function Ψ will suffice and Schrödinger's wave equation can be written

$$\frac{h}{2\pi i}\frac{\partial \Psi}{\partial t} = \left[-\frac{h^2}{8\pi^2 m}\triangle + V(x, y, z, t)\right]\Psi \tag{1}$$

or,

$$\frac{h}{2\pi i}\frac{\partial \Psi}{\partial t} = H\Psi, \quad \text{where} \quad H = -\frac{h^2}{8\pi^2 m}\triangle + V. \tag{2}$$

If, however, we wish to take into account relativity corrections, then we must use the Klein-Gordon equation which in the absence of a field may be written

$$\Box\Psi + \frac{8\pi^2}{h^2}m_0^2 c^2 \Psi = 0. \tag{3}$$

For a charged particle moving in an electromagnetic field defined by the potentials **A** and V, this equation must be generalised in a way which is now well known.

For a particle with spin $h/4\pi$, for example the electron, it is assumed that the function Ψ has four components Ψ_k ($k = 1, 2, 3, 4$) which satisfy Dirac's system of four equations. In field-free space, these equations are of the form:

$$\frac{h}{2\pi i}\frac{\partial \Psi_k}{\partial t} = \left(\alpha_1 \frac{\partial}{\partial x} + \alpha_2 \frac{\partial}{\partial y} + \alpha_3 \frac{\partial}{\partial z} + \alpha_4 m_0 c\right)\Psi_k \quad (k = 1, 2, 3, 4), \quad (4)$$

where $\alpha_1, \alpha_2, \ldots$ are the Dirac matrix operators. These equations can also be generalised in a well-known manner to the case of a finite electromagnetic field.

For particles with spin $nh/4\pi$ and $n = 2, 3 \ldots$, a wave function Ψ with more than four components, must be used, and the partial differential equations are more complicated than those of Dirac. These equations are well-known in the general theory of particles with spin.

All these wave equations have a common property of great importance. They enable us to define two real quantities, namely, a density ρ and a flux $\rho \mathbf{v}$, which can be considered as being characteristic of a hypothetical conservative fluid which is associated with the propagation of the wave. This fluid is conservative because, by virtue of the equation or equations of propagation, the quantities ρ and $\rho \mathbf{v}$ satisfy the continuity equation

$$\frac{\partial \rho}{\partial t} + \text{div}\,(\rho \mathbf{v}) = 0, \quad (5)$$

which ensures that the quantity $\int \rho d\tau$ is independent of time. It is this fact which enables us to "normalise" ρ by putting $\int \rho d\tau = 1$.

In the case of Schrödinger's equation we have

$$\rho = \Psi \cdot \Psi^* = |\Psi|^2; \quad \rho \mathbf{v} = \frac{h}{4\pi i m}(\Psi\,\text{grad}\,\Psi^* - \Psi^*\,\text{grad}\,\Psi), \quad (6)$$

where the asterisk denotes a complex conjugate. If we put $\Psi = a e^{\frac{2\pi i}{h}\varphi}$ with real a and φ, then we can also write

$$\rho = a^2; \quad \rho \mathbf{v} = -\frac{a^2}{m}\,\text{grad}\,\varphi; \quad \mathbf{v} = -\frac{1}{m}\,\text{grad}\,\varphi. \quad (7)$$

Analogous expressions can be found for ρ and $\rho\mathbf{v}$ in the case of the Klein–Gordon equation, the Dirac equation, and the equations which are suitable for particles with spin greater than $h/4\pi$. In each case ρ and $\rho\mathbf{v}$ are real and satisfy the continuity equation. However, for particles with spin greater than $h/4\pi$, it is found that ρ can assume locally negative values, which is not satisfactory for a density which, as we shall see, is interpreted as a localisation probability *.

We note that the wave function Ψ (or the components of this wave function) is complex, because the wave equation contains $i = \sqrt{-1}$. This cause little inconvenience when the wave function is used in a simple mathematical form to calculate the various probabilities. In a more realistic interpretation of the wave of wave mechanics, such as that considered in the Theory of the Double Solution, the question must be re-examined but it is evident that one can always arrive at a real equation which gives real solutions, for example, by taking a and φ to be real in the solution of Schrödinger's equation $\Psi = a e^{\frac{2\pi i}{h}\varphi}$.

2 – Localisation Principle and the Principle of Spectral Resolution

Not long after the publication of Schrödinger's original papers in the Spring of 1926, an effort was made to understand the nature and uses of the wave function Ψ. It was Max Born who in 1926–1927, was the first to discover the probabilistic interpretation of the wave function Ψ. I have generally presented my own views in the form of two distinct principles: the principle of localisation and the principle of spectral resolution. The latter was used first of all in its application to energy and momentum but was later generalised.

The localisation (or interference) principle states that the probability of finding a particle at a particular point is given by the density ρ for the species of particles under consideration. More precisely, the probability that the particle will appear at a time t in an element of volume $d\tau$ (thus giving rise to an observable phenomenon by a chain reaction process in the vicinity of $d\tau$) is given by $\rho(x, y, z, t)d\tau$. In the case of Schrödinger's equation, this probability may be written

$$|\Psi(x, y, z, t)|^2 \, d\tau = a^2(x, y, z, t) \, d\tau.$$

---

* Cf. Appendix at the end of this monograph.

This is identical with the expression which gives the intensity of light in terms of the square of the amplitude in the classical theory of light. If we can accept that light is composed of photons, we can say that the probability of the presence of a photon in a light wave is $a^2 d\tau$ and thus arrive at the connection between classical wave theory and wave mechanics.

When we come to study the Theory of the Double Solution, we shall see that the continuity equation (5) leads us quite naturally to a picture of the localisation principle. But, even in the usual interpretation, it is the continuity equation which enables us to normalise the wave function by putting $\int \rho d\tau = 1$ ($\int |\Psi^2| d\tau = 1$ in the case of the Schrödinger equation), and thus define ρ as being an absolute probability obeying the principle of total probabilities, according to which the total probability of all possible locations of the particle should be equal to 1. We note that in assuming that the physicist is correct in normalising the wave function Ψ, that is to say by assigning to it an arbitrarily chosen amplitude, we accept the subjective nature of this wave.

We now pass on to the principle of spectral resolution, which we shall first of all state for the energy and momentum of a free particle. We have known since the beginning of wave mechanics that we must associate with the uniform rectilinear motion of a free particle a monochromatic plane wave of the form

$$\Psi = A e^{\frac{2\pi i}{h}(Et - p_x x - p_y y - p_z z)}, \quad \text{where} \quad E = \frac{1}{2m}(p_x^2 + p_y^2 + p_z^2), \tag{8}$$

in the case of the non-relativistic Schrödinger wave equation, and

$$\Psi = A e^{\frac{2\pi i}{h}(Wt - p_x x - p_y y - p_z z)}, \quad \text{where} \quad \frac{W^2}{c^2} = m_0^2 c^2 + p^2, \tag{9}$$

in the case of relativistic wave equations.

Since the assumed wave equations are always linear, the general form of the wave function will be

$$\Psi = \sum_i c_i a_i e^{\frac{2\pi i}{h}(E_i t - p_{ix} x - p_{iy} y - p_{iz} z)}, \tag{10}$$

where a_i is the normalised amplitude of the i-th Fourier component.

The wave function (10) is valid in the non-relativistic case (with discontinuous spectrum) and there is a similar formula for the relativistic case. The constants c_i can be complex.

The principle of spectral resolution states that the result of a measurement of the momentum of the particle will be one of the $\mathbf{p}_i$ with a probability equal to $|c_i|^2$. If the Ψ-function has been normalised, Parseval's Theorem shows that $\sum_i |c_i|^2 = |\Psi|^2 = 1$, so that $|c_i|^2$ gives directly the absolute probability in accordance with the principle of total probabilities.

Here again, it is easy to draw a parallel with classical optics. Thus, consider a train of *light* waves

$$\Psi = \sum_i c_i \, e^{2\pi i \left(\nu_i t - \frac{\alpha_i x + \beta_i y + \gamma_i z}{\lambda_i}\right)}$$

formed by the superposition of monochromatic plane waves. In order to measure the momentum $h\nu/c = h/\lambda$ of a photon associated with this wave train, the wave train must be directed on to a grating or a prism which disperses the radiation so that different wavelengths are concentrated in different directions. This is equivalent to a Fourier analysis of the original wave train on a plane at a certain distance from the dispersing element. A wavelength λ_i, and therefore a momentum h/λ_i, can be assigned to a photon once it has been observed at a given point on this plane. It is easy to see that the probability of the value λ_i is directly proportional to $|c_i|^2$.

This can be stated more precisely as follows. The effect of the prism or grating is ultimately to effect a spatial separation of the various Fourier components of the incident wave. It is found that the phase relationships between the components are modified as a consequence of their separation. In each separate region occupied by one of the component waves, the probability of localisation of the particle is always given by $|\Psi|^2$; but since in the i-th region, where Ψ reduces to

$$c_i a_i \, e^{2\pi i \left(\nu_i t - \frac{\mathbf{n}_i \cdot \mathbf{r}}{\lambda_i}\right)},$$

we have

$$\int |\Psi|^2 \, d\tau = \int |c_i|^2 \, |a_i|^2 \, d\tau = |c_i|^2$$

(the amplitude is assumed to be normalised), we can see that the probability of assigning the wavelength λ_i to the photon after the measurement, is equal to $|c_i|^2$. This shows, in the particular example of the photon, that the measurement of the energy and momentum is always made by means of a device which *modifies the phase relationships* between the components of the Fourier series given by (8) or (9).

If the spectrum of the Ψ-wave is continuous, we can replace (8) by

$$\Psi = \int c(\mathbf{p}) \, e^{\frac{2\pi i}{h}(Et - p_x x - p_y y - p_z z)} \, d\mathbf{p}, \tag{11}$$

where $d\mathbf{p} = dp_x dp_y dp_z$ and E is expressed as a function of p_x, p_y and p_z. The probability that the measurement of the momentum of the particle will result in a value lying in the interval $dp_x dp_y dp_z$ is given by $|c(\mathbf{p})|^2 d\mathbf{p}$. Again, we have that $\int |c(\mathbf{p})|^2 d\mathbf{p} = 1$ if Ψ is normalised. We shall not dwell on the difficulties raised by a rigorous examination of the case of continuous spectra.

Instead of considering a free particle, we can consider a particle in a linear oscillator, in a hydrogen atom, in a rectangular box, and so on. Here, we have to solve a quantization problem, i.e. determine the frequencies of the stationary Ψ-waves corresponding to the form assumed for the wave equation, and to the boundary conditions set by the problem under consideration; the problem is in fact an eigenvalue problem. Thus, a set of eigenvalues ν_i will be found for the frequency (i.e. values of $E = h\nu_i$ for the energy), and the wave function representing the stationary wave of frequency ν_i (eigenfunction) will be of the form

$$\varphi_i(x, y, z, t) = a_i(x, y, z) e^{2\pi i \nu_i t} = a_i(x, y, z) e^{\frac{2\pi i}{h} E_i t}, \tag{12}$$

where a_i are normalised. The general solution of the wave equation, subject to the boundary conditions, is then

$$\Psi(x, y, z, t) = \sum_i c_i a_i(x, y, z) e^{\frac{2\pi i}{h} E_i t}.$$

The principle of spectral resolution now tells us that to each stationary state of the particle there corresponds an energy E_i, and that this energy occurs with the probability $|c_i|^2$. Thus, the values of the energy are restricted to the E_i and this constitutes the "quantization" of the energy which can be considered as a simple consequence of the principle of spectral resolution.

The principle of spectral resolution can be generalised to cases other than those of energy and momentum. It can in fact be applied to any measurable mechanical variable associated with the particle (such as the components of the moment of momentum, etc.), to an operator A which can contain the space variables, to the derivatives with respect to these variables, and even to the time t. The eigenvalue equation can be written in the form $A\varphi = \alpha\varphi$ were α is a constant and the eigenvalues α_i and eigenfunctions φ_i are determined by the boundary conditions. If the wave function has the general form $\Psi = \sum_i c_i \varphi_i$, where the φ_i are normalised, the probability that a single observation will enable the value α_i to be assigned to the variable A will be $|c_i|^2$. We shall not dwell any longer on the principle of spectral resolution, which is a classical topic.

3 – Can the Two Principles Stated above be Unified?

We have stated the laws of probability of wave mechanics by distinguishing the principle of localisation from the principle of spectral resolution. We have thus isolated the verification of localisation from the measurement of other physical quantities. Is it possible to absorb the principle of localisation into a generalised form of the principle of spectral resolution? Formally, this can be done in the following way. By using the well-known Dirac δ-function, the localisation of a particle at a point in space defined by a position vector $\mathbf{R}_0$ can be represented by the eigenfunction $\delta(\mathbf{R}-\mathbf{R}_0)$ corresponding to the eigenvalue $\mathbf{R}_0$. In view of the properties of the Dirac δ-function the explicit formula can then be written

$$\Psi(\mathbf{R}) = \int \Psi(\mathbf{R}_0)\,\delta(\mathbf{R}-\mathbf{R}_0)\,d\tau, \tag{13}$$

where $d\tau = d\mathbf{R}_0 = dx_0 dy_0 dz_0$. In applying the principle of generalised spectral resolution, it can be stated that the probability of a particle appearing in an element of volume $d\tau$ in the vicinity of the point $\mathbf{R}_0$ is $|\Psi(\mathbf{R}_0)|^2 d\tau$. Thus, as a result of this elegant mathematical argument, the principle of localisation appears to have been absorbed into the generalised principle of spectral resolution.

In recent years I have been led to suspect that this elegant demonstration, which is the basis of the transformation theory, is rather deceptive and gives a distorted view of physics. Whilst the measurement of dynamic variables such as momentum of a particle requires the physical separation of the components of the wave function (with the attendant modification of the phase relationships) and the subsequent verification of an observable localisation in one of the separated wave trains, the position of a particle cannot be completely determined by isolating the various infinitesimal components of the wave [which correspond to the eigenfunctions $\delta(\mathbf{R}-\mathbf{R_0})$] and then verifying its presence in one of them. The breakdown of the wave into infinitely small elements is physically impracticable and would not result in anything which could be observed. The observable localisation is produced only when the particle is in very close proximity to another micro-particle and when as a result of interaction, a process is initiated which is at first very localised, but after amplification by a chain reaction, produces an observable effect.

Particle localisation occurs as a result of the very passage of the particle through matter, without any causative device being set up by the physicist. Localisation is thus something much simpler and fundamental than the measurement of momentum by means of a suitable device, or by a collision with conservation of energy and momentum. The choice of the density ρ as the probability of the presence of the particle is justified by the continuity equation (5) and not by equation (13), which is only a tautology, for it simply expresses the fact that $\Psi(\mathbf{R})$ is equal to $\Psi(\mathbf{R_0})$ at the point $\mathbf{R} = \mathbf{R_0}$.

Thus, the principle of localisation seems to me to be in reality independent of the principle of spectral resolution, for the same reason that observable localisation processes play the fundamental role in microphysics.

4 – *Uncertainty Relationships*

In concluding this summary of the usual formalism of wave mechanics, mention must be made of the Heisenberg uncertainty relationships. These equations are, mathematically, a consequence of the laws of probability for two canonically conjugate variables such as x and p_x, and the laws themselves follow from the localisation principle and from the principle of spectral resolution.

One method of deducing the uncertainty relationships is based on the use of the Fourier integral. Thus, consider a wave train which, for example, at time $t = 0$, has a Fourier integral of the form

$$\Psi(x, y, z, 0) = \iiint c(p_x, p_y, p_z) e^{-\frac{2\pi i}{h}(p_x x + p_y y + p_z z)} dp_x dp_y dp_z. \quad (14)$$

Let us suppose that the wave train has maximum linear dimensions Δx, Δy, and Δz along the three rectangular coordinate axes. In order to represent it by a Fourier integral, it is necessary that the triple integral in (14) acts on the intervals Δp_x, Δp_y, and Δp_z (the c's being non-zero) such that

$$\Delta x \Delta p_x \geqq h; \quad \Delta y \Delta p_y \geqq h; \quad \Delta z \Delta p_z \geqq h. \quad (15)$$

Since, according to the localisation principle, the particle may appear at any point of the wave train, and since the measurement of its momentum can only yield values for p_x, p_y, and p_z which lie within Δp_x, Δp_y, and Δp_z respectively, it follows that Δx, Δy, Δz, Δp_x, Δp_y, Δp_z, are the "uncertainties" in the values of the coordinates and momenta, and the inequalities of equation (15) may be referred to as the "uncertainty relationships".

The uncertainty relationships can be found in another form, almost equivalent to the previous one, by starting from the concept of "standard deviation". If the values of quantity A are distributed in accordance with a certain probability law, the standard deviation σ_A is defined as the root mean square variation of A from its mean $\bar{A}$, i.e. by the formula

$$\sigma_A = \sqrt{\overline{(A - \bar{A})^2}}. \quad (16)$$

When this definition is applied to the laws of probability given for x, y, and z by the localisation principle, and for p_x, p_y, and p_z by the principle of spectral resolution, it is found that *

$$\sigma_x \sigma_{p_x} \geqq \frac{h}{4\pi}, \quad \sigma_y \sigma_{p_y} \geqq \frac{h}{4\pi}, \quad \sigma_z \sigma_{p_z} \geqq \frac{h}{4\pi}. \quad (17)$$

---

* See L. DE BROGLIE, *La quantification dans la nouvelle Méchanique*, Hermann, Paris, 1932, p. 199.

The problem now arises as to what is the significance of the uncertainty relationships? Are the $\Delta x, \ldots, \Delta p_z$ the "uncertainties" in the values of $x, \ldots, p_z$ which exist, but which we do not know exactly? Or are they a measure of the uncertainty in the values of $x, \ldots, p_z$, which arises in the process of measurement but does not exist otherwise? Do they refer to a unique situation which exists in the state of the particle symbolised by the function Ψ prior to measurement? Or do the uncertainties Δx, Δy, and Δz refer to some initial state whilst Δp_x, Δp_y, and Δp_z refer to a final state which results from the process of measuring p_x, p_y, and p_z? These very difficult questions must be discussed in order to explain the solutions given in the usual interpretation, and also those which now appear to be preferable.

5 – *The Usual Interpretation of the Formalism of Wave Mechanics*

The usual interpretation of the formalism of wave mechanics is purely probabilistic, i.e. no attempt is made to look beyond the laws of probability which we have explained, and the idea of a hidden reality, on which the laws of probability are based, is rejected. This positivistic interpretation is based on the assertion that everything which is unobservable is non-existent and should have no place in theoretical physics. This assertion seems to me to be highly debatable. The fact that a particular entity is unobservable need not preclude its use in theoretical physics if it helps in our understanding of the subject and removes undesirable paradoxes. Those who accept without reservation the purely probabilistic interpretation speak constantly of electrons, atoms and atomic nuclei, although these basic units of matter have never been directly observed. It is very difficult to deny the existence of a properly defined physical entity, just because it has not been observed.

Another idea, which is fundamental to the usual interpretation of wave mechanics is that the experimental conditions govern our understanding of the microphysical world and should always be taken into account. Bohr and Heisenberg seem to have proved that the measurement of a microphysical quantity introduces a modification of the situation which exists prior to measurement, because of the existence of the quantum of action. There is certainly no doubt that the measurement of the momentum of a particle by means of, say, a prism or a grating (in the case of photons), or by means of a collision with another particle and the application of conservation laws (for example, the Compton effect), yields a value for the momentum which may be very different from that which existed prior to measurement.

The concepts accepted *a priori* by the supporters of the purely probabilistic interpretation have led them to conclusions which, although they may be presented quite differently by different authors, can be summarised as follows.

Uncertainties such as Δx and Δp are interpreted not as simple uncertainties resulting from our partial ignorance of the exact values of position and momentum, but as the true imprecisions in these quantities. Thus, the introduction of the probability of localisation $|\Psi|^2$, which is non-zero over the whole of the wave, is not considered as being the result of our ignorance of the position of the particle within the wave, position which should exist, but which we do not know it exactly. The particle would be present throughout the entire wave in the "potential state" and it would be sharply localised only at the instant of an observable localisation. The term "potential presence" of the particle at every point of the wave, which we are almost compelled to use in order to explain this interpretation, leads to confusion: if we say that the particle *may* be present at every point of the wave, does this imply that it is at a particular but unknown point within the wave at every instant, or does it imply that the particle is mysteriously omnipresent throughout the entire wave train? The second interpretation seems to be the one adopted by supporters of the usual interpretation, although some of them maintain a prudent silence on this point.

For the uncertainty Δp_x, the question is a little different, because the measurement of momentum only tells us the value *after* the measurement and Δp_x refers to the state of the particle after measuremermt. However, the usual interpretation to be that Δp_x represents an uncertainty in the momentum in the initial state, prior to any measurement being made. This is certainly curious on the part of a theory which asserts that the process of measurement modifies radically the initial state of the microphysical system on which the measurement is performed.

Briefly, it seems that the usual interpretation regards the uncertainties Δx and Δp_x, which occur in the Heisenberg inequalities, as representing the true imprecision in the position and momentum of the particle in the state defined by the wave Ψ *before* any observable verification of the position, or before any measurement is made of the momentum.

We must consider whether this interpretation is truly satisfactory.

In the usual interpretation of wave mechanics the wave function is considered as being essentially non-objective in character, and apparently conclusive reasons are given for this. First of all, in order to be able to represent probabilities by means of $|\Psi|^2$ and $|c_i|^2$, the wave Ψ must be normalised, that is to say, its amplitude must be chosen in a quite arbitrary manner so that $\int |\Psi|^2 d\tau = 1$. Now, quite obviously, the amplitude of an objective wave cannot be chosen arbitrarily. Dirac has frequently stressed this fundamental difference between the waves of classical physics and the wave Ψ of wave mechanics. On the other hand, Heisenberg has drawn attention to what he has called the "reduction of the wave packet" by observation *. As an example, consider a wave train associated with a particle (for example a photon) which strikes a semi-transparent mirror. The initial wave train is divided into two waves, one being transmitted and the other reflected. Suppose now that an observable localisation enables us to confirm that the particle appears in one of the wave trains, say, in the transmitted wave train. Then the reflected wave train no longer offers a possible localisation and should henceforth be considered as non-existent so that the wave must be renormalised and the transmitted wave train continues alone. This reduction of the wave packet, many examples of which could be given, demonstrates clearly that the wave Ψ is not of an objective nature but is only a subjective representation of probabilities.

However, we have seen that it is very difficult not to assign an objective character to the wave of wave mechanics which is propagated, reflected, and diffracted, and which determines the phenomena of interference and diffraction of light and of particles, as well as the stationary states of quantized systems. Faced with the phenomena of interference and diffraction, a physicist with no theoretically preconceived ideas is compelled to believe that he is dealing with the propagation of real waves and not with a simple representation of a probability which exists only in his mind.

Thus, we are once again faced with the riddle: the true wave of wave mechanics should be objective and able to determine physical phenomena, whilst the wave function Ψ is only a representation of probability of a subjective nature. We have already shown how the Theory of the Double

---

* "Wave packet" simply implies "wave train". The term "probability packet" is also sometimes used.

Solution endeavours to resolve this riddle and we shall return to it shortly.

The usual interpretation of wave mechanics, which denies the permanent localisation of particles, does not explain adequately the appearance of the observable localisations upon which all our knowledge of the microphysical world is based. Many authors have endeavoured to explain localisation and microphysical measurements in terms of the effect of the measuring device, but we have seen that measuring apparatus, in the normal sense of the word, plays very little part in this problem. Certainly, in order to understand how an observation is carried out, it is essential to recognise the measuring device and its manner of operation, but it is not necessarily the device which creates the final result. Moreover, in the phenomenon of observable particle localisation there is, strictly speaking, no measuring device. Some authors, who are undoubtedly dissatisfied with the usual explanation of the effect of the device, have supported the von Neumann theory in which the result of the measurement is created by the consciousness of the observer. This point of view has led them to completely unacceptable conclusions *. It appears that none of the explanations offered so far are satisfactory and that the question must be taken up on another basis, as we shall show below.

The usual theory describes the stationary states of a quantized system by assigning to them the appropriate eigenfunctions, but is completely incapable of describing the rapid transitions which take place from one stationary state to another, quantum transitions which accompany the emission of radiation in Bohr's theory of the atom. A long time ago Bohr said that the description of these quantum transitions has "transcended" the framework of space–time, which is purely and simply a refusal to offer an explanation. Schrödinger also was able to say with humour: "the present theory describes minutely the stationary states, which are of no interest because nothing takes place there, but it remains silent about the intermediate states". This remark emphasises an insolvency which has, as we shall see later, a very profound significance.

---

* See Bibliography [2], p. 38 et seq.

CHAPTER III

Difficulties Raised in the Present Theory by the Hypothesis that the Particle is not Constantly Localised in Space

1 – *Einstein's Objection*

A general criticism of the usual interpretation of wave mechanics is that it constitutes a verbose refusal to provide a real explanation and this is, it seems to me, quite contrary to the principles of sound scientific method. This applies to the concept of complementarity, even when presented in an acceptable form, to the assertion of a potential presence of the particle in an extended region of space, to the "transcendent" nature attributed to quantum transitions, and so on. Leaving aside these general objections however, we shall concentrate our attention on the difficulties raised by the hypothesis that, apart from its observable manifestations, the particle is not constantly localised in space.

To begin with, let us consider the objection presented by Einstein at the Solvay Council in 1927. Einstein considered a particle falling normally

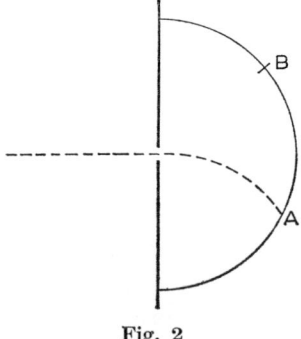

Fig. 2

on a plane screen pierced with a small hole. A photographic film is placed behind the screen and is in the form of a hemisphere of large radius, centred on the hole.

If the dimensions of the hole are small compared with the wavelength of the wave associated with the particle, then the wave will be diffracted and will be distributed over the entire hemispherical film, since the hole will now act as a small centrally located point source. If at some time t a photographic record (constituting an observable particle localisation) is obtained at a point A on the film, its interpretation will be very different depending on whether a trajectory is assigned to the particle, or whether the usual wave mechanical theory is adopted.

The dotted line in Fig. 2 represents diagrammatically the classical trajectory. This trajectory must intersect the screen, but until the localisation is established, the trajectory followed will not be known, and we must assign a non-zero probability (equal to $|\Psi|^2$) to the presence of the particle at every point. This probability expresses our lack of knowledge of the path followed by the particle. As soon as the particle appears at the point A, we know that its trajectory, *whatever may have been its shape*, has converged on this point, and the probability of finding this particle at another point on the screen immediately becomes zero. All this is very clear.

However, in the usual interpretation of wave mechanics, there is no trajectory associated with the hemispherical wave on the right of the screen. So long as the localisation at point A has not taken place, the particle must be considered as present in the potential state over the whole surface of the screen with the probability $|\Psi|^2$. As soon as the particle appears at the point A, the probability of finding it at another point on the screen vanishes since, by hypothesis, there is only one particle associated with the wave. The interpretation of this fact, which is quite simple when the existence of a trajectory is accepted, here becomes very mysterious. In fact, the principle of relativity asserts that no signal can be propagated with a velocity greater than the velocity of light c, so that it is impossible to accept that the localisation which has taken place at point A is able to signal instantaneously its presence to all points on the screen (which can have very large dimensions). Moreover, it is not clear why the emission of a single particle by the wave source is not able to produce more localisations on the screen. Einstein considered that the absence of any direct causative link, resulting from the existence of a trajectory joining the emission of the particle by the source to its observable localisation on the screen, constituted an insurmountable conceptual difficulty.

In order to meet this objection, we could endeavour to introduce a finite time of propagation of the wave through the screen. The whole wave is, in fact, of finite size and constitutes a train of waves having a thickness l, with leading and trailing edges. In Einstein's example, the spherical wave to the right of the screen constitutes a spherical shell which moves with the particle velocity v. This shell takes a time l/v to traverse the screen, and during this time a signal which has left point A would reach points on the screen situated at a distance less than or equal to cl/v from A. It is obvious that this does not in any way enable us to remove Einstein's objection. Moreover, for an electron, l is of the order of 1/10-th millimetre, v is 10^9 cm/sec, and cl/v is of the order of a millimetre. Clearly, the screen will generally have much greater dimensions.

Another possible approach is to regard the wave as only a probabilistic representation, such that as soon as the localisation at the point A has taken place, the probability of localisation at all other points becomes zero and there is no need for the "signal" from A to be transmitted to all other points. However, this brings us back to the difficulty of having to regard the wave as being subjective in nature.

2 – Schrödinger's Objection

Schrödinger, in a series of articles *, joined Einstein in rejecting the probabilistic interpretation of wave mechanics, and developed the framework of an objection which explicitly introduced the finite dimensions of the wave trains and which appears to me to be of great significance.

Consider two groups of nearly monochromatic waves associated with two particles 1 and 2. On close approach, the two particles interact, and

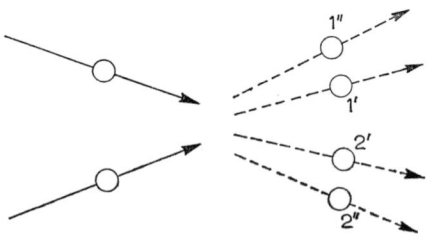

Fig. 3

---

* *Naturwissenschaften*, 23 (1935) 787, 823, 844.

the usual theory describes the possible outcome of the interaction in terms of the propagation of a Ψ wave in configuration space. The wave mechanics of configuration space (to which we shall return) now teaches us that the collision may give rise to a series of finite possible motions, all of which are compatible with the conservation of energy and momentum. Thus, either the wave train of particle 1 will ultimately follow a trajectory 1', and the wave train of particle 2 will follow the *correlated* trajectory 2'; or the wave train of particle 1 will follow the trajectory 1", and the wave train of particle 2 will follow the trajectory 2", and so on. The wave trains in the final state will thus form correlated couples, 1' with 2', 1" with 2", and so on.

Now, let us suppose that we observe a particle localisation for particle 1 in the wave train 1'. We shall then know that particle 2 is in wave train 2'. This is easily understood if the particles occupy, at any instant of time, a certain position in the physical space inside their wave trains, for now we can say that after impact particle 1 follows the trajectory 1', and particle 2 follows the trajectory 2'. The observable localisation which is produced in 1' only indicates that the particle, in following its trajectory, was in wave train 1' and thus particle 2 should be found in wave train 2'. All this is very clear but does not correspond to the usual interpretation.

The orthodox interpretation implies that the particle is *not* localised in the wave. After impact, particle 1 would be potentially present in the aggregate of wave trains 1', 1", ..., and particle 2 in the aggregate of wave trains 2', 2", ... Whilst an observable localisation is produced by particle 1 in 1', particle 2—*on which no action was exerted*—would be forced into wave train 2' to the exclusion of 2", ..., even though 2' should, at this instant, be located at a considerable distance from 1'. As Schrödinger said, "this would be magic".

We note that it can be neither the measuring device nor still less the skill of the observer which enables the observation of the particular localisations to be made. Mention is often made of the "uncontrollable reaction" exerted on the particle by the measuring device, which forces it into one or other of the possible final states. In the case of the correlated systems which we have just considered, such an interpretation appears to be quite meaningless.

3 – *Another Form of the Preceding Objections*

In a recent article in *Journal de Physique* [3], I have revived the above objections in a new form.

Suppose a particle is enclosed in a box B with impermeable walls. The associated wave Ψ is confined to the box and cannot leave it. The usual interpretation asserts that the particle is "potentially" present in the whole of the box B, with a probability $|\Psi|^2$ at each point. Let us suppose that by some process or other, for example, by inserting a partition into the box, the box B is divided into two separate parts B_1 and B_2 and that B_1 and B_2 are then transported to two very distant places, for example to Paris and Tokyo. The particle, which has not yet appeared, thus remains potentially present in the assembly of the two boxes and its wave function Ψ consists of two parts, one of which, Ψ_1, is located in B_1 and the other, Ψ_2, in B_2. The wave function is thus of the form $\Psi = c_1\Psi_1 + c_2\Psi_2$, where $|c_1|^2 + |c_2|^2 = 1$.

The probability laws of wave mechanics now tell us that if an experiment is carried out in box B_1 in Paris, which will enable the presence of the particle to be revealed in this box, the probability of this experiment giving a positive result is $|c_1|^2$, whilst the probability of it giving a negative result is $|c_2|^2$. According to the usual interpretation, this would have the following significance: since the particle is present in the assembly of the two boxes prior to the observable localisation, it would be immediately localised in box B_1 in the case of a positive result in Paris. This does not seem to me to be acceptable. The only reasonable interpretation appears to me to be that prior to the observable localisation in B_1, we know that the particle was in one of the two boxes B_1 and B_2, but we do not know in which one, and the probabilities considered in the usual wave mechanics are the consequence of this partial ignorance. If we show that the particle is in box B_1, it implies simply that it was already there prior to localisation. Thus, we now return to the clear classical concept of probability, which springs from our partial ignorance of the true situation. But, if this point of view is accepted, the description of the particle given by the customary wave function Ψ, though leading to a perfectly *exact* description of probabilities, does not give us a *complete* description of the physical reality, since the particle must have been localised prior to the observation which revealed it, and the wave function Ψ gives no information about this.

We might note here how the usual interpretation leads to a paradox in the case of experiments with a negative result. Suppose that the particle is charged, and that in the box B_2 in Tokyo a device has been installed which enables the whole of the charged particle located in the box to be drained off and in so doing to establish an observable localisation. Now, if nothing is observed, this negative result will signify that the particle is not in box B_2 and it is thus in box B_1 in Paris. But this can reasonably signify only one thing: the particle was already in Paris in box B_1 prior to the drainage experiment made in Tokyo in box B_2. Every other interpretation is absurd. How can we imagine that the simple fact of having observed *nothing* in Tokyo has been able to promote the localisation of the particle at a distance of many thousands of miles away?

4 – Renninger's Negative-Result Experiment

What we have just said leads us to mention a recent article by W. Renninger [4], who for many years has pursued a campaign in Germany against the usual interpretation of wave mechanics *. Renninger has emphasised the striking paradox to which this interpretation leads in the case of "negative" experiments.

In the example which he gives, a point source S emits particles isotropically in all directions. A screen E_1 in the form of a sector of a sphere centred on S and having a radius R_1 is covered on the inside with a

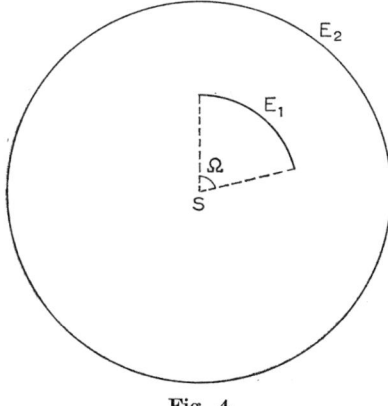

Fig. 4

---

* See especially the very remarkable article by this author in Z. *Physik*, 13b (1953) 251.

substance which indicates the arrival of a particle by a scintillation (observable particle localisation!). Another screen E_2, in the form of a complete sphere centred on S and having a radius $R_2 > R_1$, completely surrounds the screen E_1. The second sphere is also covered inside with a phosphor.

Suppose now that the screen E_1 subtends a solid angle Ω at S. The propagation of the wave emitted by S is restricted by the screen E_1, and diffraction phenomena occur at the edges of E_1. Notwithstanding the existence of these diffraction phenomena, it is obvious that a particle emitted by the source will have a probability $P_1 = \Omega/4\pi$ of producing a scintillation on E_1 and a probability $P_2 = (4\pi-\Omega)/4\pi$ of producing a scintillation on E_2. At the instant of emission by the source of a particle with a velocity v, the emission of the associated wave commences at a time $t = 0$ and lasts for a finite time τ. The emitted wave Ψ forms a spherical shell whose leading edge reaches the screen E_1 in a time $t_1 = R_1/v$ while the trailing edge reaches the same screen at the time $t_1+\tau$. If, at time $t_1+\tau$ no scintillation is produced on screen E_1, we can be certain that the scintillation will be produced on E_2. P_1 suddenly becomes zero, and P_2 becomes equal to 1. Thus, there will be a sharp change in the amplitude of the wave on the two screens and, according to the usual theory, we shall have a special case of the reduction of a probability packet. A particularly paradoxial situation will now exist, since the observer sees nothing at all on screen E_1, where nothing has happened. In this experiment the reduction of the probability packet is quite incomprehensible. It is, in fact, impossible to accept that this reduction is due to the increase of knowledge of the observer who has observed nothing, nor to a device—here screen E_1—which registered nothing.

The situation becomes clearer if we accept that the source emits a particle which remains closely associated with a wave, but which has a definite position at each instant of time and, consequently, a definite trajectory. The trajectory should be closely linked with the propagation of the wave and should be influenced by it. It can be accepted that, at least on an average*, these trajectories are straight lines starting from S, except for the immediate vicinity of the edges of screen E_1 which constitute an obstacle to propagation of the wave and give rise to diffraction, thus producing local modifications of the trajectories. On the whole, it can be

---

* The Bohm–Vigier perturbations will be considered later.

said that the number of possible trajectories emanating from S and terminating on E_1 is proportional to Ω whilst the number of trajectories emanating from S and reaching E_2, whether after a rectilinear trajectory or a trajectory which has been disturbed by diffraction at the edges of screen E_1, is proportional to $4\pi-\Omega$. We thus find the probabilities $P_1 = \Omega/4\pi$ and $P_2 = (4\pi-\Omega)/4\pi$ of arrival of the particle at either E_1 or E_2. If no scintillation is produced on E_1 in the time τ, which is the time taken by the wave to reach the sphere of radius R_1, then we may be sure that the trajectory followed by the particle is not one of those terminating on E_1. There would thus be a sudden change to $P_1 = 0$ and $P_2 = 1$. This sudden change will represent simply a change in the state of our knowledge of the trajectory of the particle. This removes the incomprehensible effect of the mind of the observer on the particle since there is no scintillation on E_1. As far as the presence of "measuring devices" is concerned, the screen E_1 is simply an obstacle to the propagation of the wave and thus influences the *possible* trajectories by stopping certain trajectories and giving rise to diffraction. This interpretation is very clear and much more comprehensible than the one based upon a mysterious effect which imposes on the particle the simple "possibility" that it might become localised on E_1. How can we possibly imagine that a possibility which has never become real would have such an effect?

Renninger has also considered the case where the inner screen E_1 is a complete sphere pierced by a small hole. Under these conditions, the portion of the wave which passes through the hole gives rise to diffraction patterns on E_2. We thus arrive at a classical deduction of the uncertainty relationships. However, the screen E_1 only acts as an obstacle to the propagation of the wave, and the interpretation of what happens when nothing is observed on E_1 becomes clear only if we assume that the particle follows a trajectory which passes through the hole and is influenced by diffraction.

5 – *An Attempt to Remove the Foregoing Difficulties*

The above difficulties may be countered by the following argument which is quite different from the usual interpretation and the Copenhagen ideas [*]. In this theory, "events" are permanently inscribed on the

---

[*] This argument appears to me to be very closely related to that recently discussed by Costa de Beauregard.

space–time continuum. When an event is observed, it is just because it has always been available in space–time. If an event has the coordinates x_0, y_0, z_0, t_0 in an observer's system of coordinates x, y, z, t, then it appears inevitably to the observer at the point x_0, y_0, z_0 of his own space at the instant when his clocks indicate the time t_0. This is an acceptable point of view which is certainly in accordance with the general principles of the Theory of Relativity. If this point of view is adopted, then observable particle localisations must be considered as predetermined events in space–time, which appear inevitably to the observer at the instant t_0 of his own time. This seems to be quite different from the point of view of the majority of the supporters of the present-day interpretation. Some adherents of the Copenhagen School even believe that microphysical phenomena are subject to a real indeterminism and that microparticles choose their future at every instant. In a book recently translated into French under the title *La physique et le secret de la vie organique* (Albin Michel, translated by André George), the physicist M. Pascual Jordan gave an explanation of his views on the problem. While stating (p. 48) that radiation can appear sometimes in the form of a wave and sometimes in the form of a particle (which, as we have seen, does not appear to be strictly so), he writes (p. 39): "Some of the laws which are known precisely to us at present, ascribe to the atom certain possible reactions which only allow it the *choice* of one or other of these possibilities." Further on (p. 53) he speaks of "the incessant intervention of independent decisions which represents the behaviour of atoms". These passages, where the word "atom" is specifically applied to all microparticles, seem to attribute to the succession of observable microphysical phenomena the character of events which are in no way determined in advance, and therefore cannot be considered as pre-existent in space–time.

The argument which we have been describing above leads to completely different conclusions. If events are pre-determined in space–time, they should appear inevitably at their particular time, and there is no indefiniteness. It is a kind of "fatalism", but it is not a causative determinism. Events recorded in space–time would not be interlinked by causative relationships, but they should, nevertheless, be distributed in space–time in a random manner which should conform to the laws of probability of wave mechanics. Einstein's objection, and other similar objections, would thus be removed quite unexpectedly. Let us take Einstein's objection: a particle which has been emitted by the source will manifest its presence on the hemispherical screen by a scintillation (observable localisation) at a point A at the instant of time t simply because the point in space–time defined by A and t, i.e. the "scintillation event", was permanently available, and if no other scintillations are produced at other points of the screen, it is so because no other predetermined event of this kind existed in space–time. If a new particle is emitted by the source, a new scintillation will appear on the screen at a point B at time t'. When a large number of particles has been emitted by the source, the points at which scintillations are produced on the screen will be distributed according to the laws of probability of wave mechanics, simply because the corresponding point-events are so distributed in space–time.

Einstein's objection thus seems to be removed, and all the other objections to the usual interpretation which we have studied above will be similarly removed.

On reflection, this succession of observable facts in the microphysical world may be found to be quite difficult to accept. As we have seen, in this theory point-events are assumed to have been permanently dispersed in space–time and correspond to observable microphysical events, i.e. observable particle localisations. This dispersion, whilst random in nature and without causative connection, does nevertheless follow certain laws of probability. The state of affairs now becomes most improbable. The laws of probability are derived from the propagation of a wave which is evolved in space–time according to some propagation equation, and which can be reflected at mirrors, diffracted by an obstacle, and so on. Is this wave purely imaginary and unreal, or is it an objective and real wave? We always return to the same problem. If the wave is purely imaginary and is only a means of calculating probabilities, how can it be subject to physical conditions for propagation, reflection, diffraction, and so on, which involve boundary conditions due to the presence of obstacles? If the wave is factual, it should be represented in the Theory of Relativity by a field distributed in space–time and associated with, for example, the energy–momentum tensor. Since, however, we are concerned with the wave Ψ of normal wave mechanics, this field must be uniform and cannot exhibit regions of high field concentration, representing the existence of events of a corpuscular nature. The riddle thus remains entirely unresolved.

Actually, the concept which has just been explained amounts to the

34 DIFFICULTIES RAISED BY NON-LOCALISATION

acceptance of a kind of pre-established harmony between the random distribution in space–time of the point-events representing particle localisations, and the propagation of the wave, the real or unreal nature of which still remains very obscure. The assumption of a pre-established harmony always amounts to, more or less, a rejection of an explanation, which is the same as saying: "It is so *because* it is so." This attitude, which is almost too convenient, appears to me to be quite contrary to the true scientific spirit.

Again we can ask: Why does the emission of a single particle by the source give rise to a single scintillation on the screen? The only answer to this is again: "It is a pre-established harmony—it is so because it is so."

I now propose to compare the theory discussed in this section with the argument which was surely in Einstein's thoughts and which, on careful reflection over a number of years, I consider to be preferable.

6 – *Einstein's Argument*

We can illustrate the argument developed in the preceding chapter in the following manner.

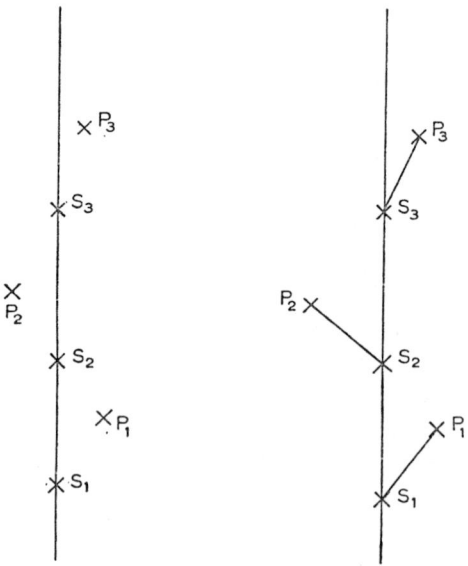

Fig. 5

Consider a source of particles and a Galilean reference system linked with this source.

The world line of the source can be represented by a vertical line which is the axis of the proper time of the source. The point-events S_1, S_2, S_3, ..., represent successive emissions of particles by the source at times t_1, t_2, t_3, ... After every emission, the particle gives rise to an observable localisation on the screen (scintillation): these localisations are represented by the point-events P_1, P_2, P_3, ... Each of the P_i must be located inside a cone of light with vertex S_i (because the velocity of a signal cannot exceed the velocity of light c). In the hypothesis developed above, which postulates a pre-established harmony which dispenses with the entire explanation, the points P_i are in one-to-one correspondence with the points S_i, but are distributed relative to S_i according to the laws of probability of wave mechanics without any causative connection.

On the other hand, for Einstein, a causative connection must exist between every S_i and P_i because there is a world line (or more precisely an infinitely thin world tube) representing the motion of the particle from the source up to the point on the screen where it becomes localised*.

However, in order to obtain a truly physical picture, it is necessary to assign an objective nature to the wave associated with the particle, which seems also required by the properties of propagation, reflection, interference and diffraction. A difficulty now arises: if we imagine that the associated wave possesses the homogeneity of the usual wave Ψ, how can we explain the sudden concentration of energy which is produced when the particle is localised? We thus arrive at the idea that a homogeneous wave of the type normally considered in wave mechanics can provide a *precise* representation of the probability of localisation, but that it does not give us a *complete* picture of the wave–particle duality. The picture of the wave propagating in space must be augmented by the addition of the world tube of the particle, which is *hidden* but establishes a causative connection between its successive localisations. The shape of this tube (i.e. the motion of the particle) must be linked in some way with propagation of the wave. Unless this is so, the wave

---

* In Fig. 5, the world lines $S_i P_i$ are drawn, for the sake of simplicity, as straight lines. Actually, they may be curved or even, if we take into account interactions of the particle with the subquantized medium, they may be infinitely broken lines. However, it is essential that there be a continuous world line joining each S_i to the corresponding P_i. The same remark also applies to the trajectory shown in Fig. 2.

gives no information about the possible localisations of the particle. The bond between wave and particle must be so intimate that we must believe the world tube is a part of the *structure* of the wave. We are thus approaching the ideas which had originally led to the concept of the Theory of the Double Solution. Nevertheless, there is a serious difficulty. If the true wave associated with the particle is a factual objective wave, containing within its structure a permanent inhomogeneity which is the particle, then in the strict meaning of the word, it cannot be completely identified with the usual wave Ψ of wave mechanics. The latter wave is undoubtedly subjective in character and has a probability representation which is normalisable at will and involves, as we have seen, the sudden "reduction of the wave packet", the non-objective character of which is obvious. Nevertheless, these two waves must be intimately related, and the problem now arises as to what is the form of this relation.

Having completed our examination of the various possible interpretations of wave mechanics, we are now ready to consider the principles of the Theory of the Double Solution.

CHAPTER IV

Summary Explanation of the Theory of the Double Solution

1 – Basic Concepts *

In my early work on wave mechanics, I adopted the view that it was essential to retain the concept of a physical reality independent of the observer, as in classical physics, and tried to obtain a clear representation of physical processes within the framework of space and time. This led me to a search for a synthetic view of the wave–particle duality which would be compatible with my early ideas (Notes and Thesis on Wave Mechanics, 1923–1924), and which had just been confirmed in a remarkable manner (Schrödinger's papers in 1926, discovery of electron diffraction in 1927). Following the work of Mie and Einstein, I was endeavouring to represent the particle as a kind of local accident, a singularity within an extensive wave phenomenon. Accordingly I introduced the u-solution of Schrödinger's wave equation, which differed from the normal Ψ-solution and permitted the inclusion of a singularity. The idea of a particle "incorporated" in an extensive wave field, and closely linked with the evolution of the field, seemed to me to be of great significance. It appeared to me to permit an understanding of how the particle is localised and how its motion could still be influenced by the presence of obstacles at a distance from its trajectory, which is necessary for the localisation of the particle and the simultaneous existence of interference phenomena. Moreover, we saw, at the end of the last chapter, that the re-introduction of a world line which represents the motion of the particle within the wave, permits a much clearer representation of microphysical phenomena.

At the same time, the probabilistic interpretation of the Ψ-wave, originally introduced by Max Born, seemed to me to be preserved. Thus, whilst the u-wave would be the true description of microparticles, the

---

* For a more complete account of these ideas, see [1], [2] and [3].

Ψ-wave would be a fictitious wave with a subjective character but capable of supplying precise statistical information and still linked in some way with the u-wave.

As a result of my initial work on wave mechanics, I considered that particular importance should be assigned to the phase of the wave associated with the particle. It was essentially the agreement between the phase of the particle, regarded as a little clock, and the accompanying wave, which led me to the two formulae $W = h\nu$, $p = h/\lambda$. The frequency and wavelength, which are contained in the phase, establish the relationship between propagation of the wave and the motion of the particle. This led me to write the usual wave function in the form

$$\Psi = a\,e^{\frac{2\pi i}{h}\varphi} \tag{1}$$

where a and φ are real, and are the amplitude and phase respectively (in the geometrical optics approximation, the phase φ corresponds to the Jacobian function S). The amplitude a appeared to me to be of statistical rather than objective significance.

The probability $|\Psi|^2$ of observable particle localisation, seemed to me to be the most important of the probabilities contained in the probabilistic interpretation of wave mechanics because, in my opinion, it indicated the possibility that the particle could be found at a point independently of any measurement. The other probabilities, such as $|c(\mathbf{p})|^2$ for the value $\mathbf{p}$ of the momentum should, in my opinion, be of less immediate significance: they would only be valid *after* the real u-wave in which the particle would be incorporated has been affected by a measuring device. The importance of these ideas has already been indicated and we shall return to them when criticising the currently accepted theory of transformations.

Having arrived at these general ideas, I postulated the Principle of the Double Solution which states: "To every regular solution $\Psi = a\,e^{\frac{2\pi i}{h}\varphi}$ of the wave equation of wave mechanics, there corresponds a solution of the form

$$u = f\,e^{\frac{2\pi i}{h}\varphi}, \tag{2}$$

having the same phase φ, but with an amplitude f involving a point singularity which is, in general, in motion."

BASIC CONCEPTS

The u-wave, which is the true physical representation of the particle, describes the wave propagation quite well when it is supplemented by the world tube representing the motion of the singularity (particle). The relationship between this representation and the previous ideas is quite clear.

In 1927, by applying the Principle of the Double Solution to the Schrödinger and Klein–Gordon equations, I was able to establish the following results:

1) It is possible to associate a solution having a point singularity with the monochromatic plane-wave solution

$$\Psi = a\, e^{\frac{2\pi i}{h}(Wt - \mathbf{p}\cdot\mathbf{r})}$$

of the Klein–Gordon equation in the absence of fields. In the usual interpretation the plane-wave solution represents uniform linear motion of a particle with energy W and momentum $\mathbf{p}$. The singularity in the associated solution is then found to move in the direction of $\mathbf{r}$ with the velocity $v = pc^2/W$ and the same phase as the plane wave. This shows that the Principle of the Double Solution is well verified in this particular case.

2) If there are two solutions of the Klein–Gordon equation, Ψ and u, one with a continuous amplitude and the other involving a variable point singularity *with the same phase* φ, the singularity of u will be displaced in space with the instantaneous velocity

$$\mathbf{v} = -c^2 \frac{\operatorname{grad} \varphi + \dfrac{\varepsilon}{c}\mathbf{A}}{\dfrac{\partial \varphi}{\partial t} - \varepsilon V}, \qquad (3)$$

where ε is the electrical charge, and V and $\mathbf{A}$ are the scalar and vector potentials of the electromagnetic field acting on the particle.

The expression given by (3) is the "guidance formula" which, if we can neglect relativistic corrections and assume that there is no magnetic field (i.e. let $\partial\varphi/\partial t \simeq m_0 c^2 + \varepsilon V$ and $\mathbf{A} = 0$), takes the simple form

$$\mathbf{v} = -\frac{1}{m}\operatorname{grad}\varphi, \qquad (4)$$

which corresponds to the Schrödinger equation and can be derived directly from it. If, moreover, the propagation of the wave takes place approximately as in geometrical optics, then we can put $\varphi \simeq S$, where S is the Jacobian function, and equation (4) becomes identical with the classical formula $m\mathbf{v} = -\mathrm{grad}\, S$ of the Hamilton–Jacobi Theory.

In my opinion, the true significance of the guidance formula lies in the fact that the particle, which is regarded as a small clock, moves so that it remains constantly in phase with the wave which surrounds it.

3) The motion of the particle is the same as if it were subjected to a "quantum force" equal to $-\mathrm{grad}\, Q$ in addition to the classical force originating from external potentials, where Q is a "quantum potential" unknown in classical theory which, on a non-relativistic approximation of Schrödinger's equation, can be written simply as

$$Q = -\frac{h^2}{8\pi^2 m}\left(\frac{\triangle f}{f}\right) = -\frac{h^2}{8\pi^2 m}\left(\frac{\triangle a}{a}\right), \qquad (5)$$

where the quantities in brackets are taken from the point where the particle is to be found at time t.

Moreover, the guidance formula and the quantum potential enables us to express the equations of motion of the particle, incorporated as a singularity in the u-wave, in a Lagrangian form.

All these problems, and in particular, the proofs of the guidance formula, are discussed in papers [1] and [2] of the Bibliography.

A particularly interesting interpretation can be given to the guidance formula by the introduction of hydrodynamic flow defined by a density ρ and a flux $\rho\mathbf{v}$ which can always be associated, as we have seen in Chapter II (§ 1), with the propagation of the Ψ-wave. In the case of Schrödinger's equation, equation (7) of Chapter II shows that the velocity v of the fluid associated with propagation of the wave is given by equation (4) above, i.e. it is the same as the velocity defined by the guidance formula. In the case of the Klein–Gordon equation, we find that

$$\rho = \frac{1}{m_0}\frac{\partial \varphi}{\partial t} - \frac{\varepsilon}{m_0 c^2}Va^2, \qquad \rho\mathbf{v} = \frac{1}{m_0}a^2\,\mathrm{grad}\,\varphi - \frac{\varepsilon}{m_0 c}a^2\mathbf{A} \qquad (6)$$

and, dividing the second expression by the first, the guidance formula is obtained in a relativistic form given by equation (3). For the Dirac equation, the same method leads to a guidance formula which is a little more complicated than will be found in [1]. Finally, the guidance formula can be used to show that the particle follows a streamline of the hydrodynamic flow associated with the wave, and has the local velocity of this flow.

In the form in which we have just given it, the guidance theory already provides us with a better visualisation of the relationship between the u-wave and Ψ-waves which we shall consider in greater detail later. These two very different waves should have the same streamlines. Thus, the Ψ-wave represents, just as well as the u-wave, the aggregate of the possible motions of the particle, but it lacks the essential association of the particle with *one* of the streamlines. This accounts for the fact that while the Ψ-wave can give a precise statistical picture of the possible particle localisations, it cannot provide a complete description of microphysical reality. We are thus driven again to the conclusion which we reached at the end of the preceding chapter and which is in agreement with the opinion which Einstein always maintained.

2 – *Introduction of a Random Element into the Foregoing Concepts*

The ideas which I have just explained, correspond to the picture of the particle in the original form of my Theory of the Double Solution in 1926–1927. In consequence of a paper by Bohm and Vigier—of which I shall speak presently—I have increasingly felt that this picture must be augmented by the introduction of the "subquantum medium".

This need becomes apparent when we attempt to justify, in the Theory of the Double Solution, the identification of the density ρ with the probability that the particle will be found at a particular point. I had attempted to do this earlier by starting from the equation of continuity

$$\frac{\partial \rho}{\partial t} + \operatorname{div} \rho \mathbf{v} = 0, \tag{7}$$

which is always valid for hydrodynamic flow associated with wave propagation. As we have just seen, in the Theory of the Double Solution, where the velocity of the particle is defined by the guidance formula, the completely localised particle follows one of the hydrodynamic flow lines. From this result I naturally concluded that if we do not know that the particle follows the trajectories defined by the guidance formula, then the probability of the presence of the particle in an element $d\tau$ of space is equal to $\rho d\tau$. On the basis of equation (7), which states that the quantity $\rho d\tau$ conserves its value when the element $d\tau$ is carried away by the flow along a tube of stream-lines, I had given several reasons which appeared to make my conclusion probable.

However, no matter how natural this conclusion may have appeared, it must not be regarded as being rigorous for the following reason. Consider a small element $d\tau$ of a hypothetical fluid in hydrodynamic flow. This element can be considered as containing $\rho d\tau$ molecules of this fluid. In due course, the element of volume containing these molecules sweeps through a small "stream-tube" formed by the stream-lines which it contains. As mentioned previously, the equation of continuity (7) can thus be interpreted by saying that the quantity $\rho d\tau$ remains constant along the stream-tube although ρ and $d\tau$ could vary individually. However since there is no reason for accepting *a priori* that a similar stream-tube would fill the entire physical space, the conservation of $\rho d\tau$ along a stream-tube does not warrant the firm conclusion that $\rho d\tau$ is the probability of presence of the particle in the element $d\tau$ when we do not know along which of the stream-lines it passes.

The difficulty encountered here is analogous (without being wholly identical) to that encountered in the kinetic theory of gases where the motion of representative points is considered in phase space. In the classical kinetic theory, the difficulty may be overcome by introducing the idea that the molecules, because of their constant mutual collisions, are subjected to constant random perturbations corresponding to what Boltzmann described as "molecular chaos". We can try to introduce a similar concept into the Theory of the Double Solution. This was done in an important paper by Bohm and Vigier [5]. According to them, the particle is subjected to constant random perturbations resulting from its continual interaction with a hidden, subjacent "subquantum medium", which is a kind of "hidden thermostat".

If we accept this hypothesis and assume that the random perturbations experienced by the particle can be represented by the momentary appearance in the wave equation of small perturbation terms, the equation of continuity will still remain valid, even during the perturbations, and the quantity $\rho d\tau$ will remain constant along a stream-tube, even in the disturbed portions where it no longer follows the regular hydrodynamic flow. Thus, similar elements $d\tau$ will pass constantly from one undisturbed stream-tube to another neighbouring undisturbed stream-tube and $\rho d\tau$ will be conserved. Therefore, we can consider that the element $d\tau$ sweeps successively through the undisturbed stream-tubes, $\rho d\tau$ is conserved, and the element can traverse the entire physical space accessible to the particle.

Thus, the motion of the particle comprises a kind of Brownian motion *superimposed* on the regular motion defined by the guidance formula. This appears to be completely analogous to what happens in a constant flow of a fluid in which the assembly of streamlines considered in hydrodynamics form the aggregate of possible trajectories of molecules in undisturbed motion, but where in reality every molecule is in continuous Brownian motion which is of thermal origin and which causes it to pass constantly from one streamline to another. The particle thus appears as a kind of granule carried by the flow associated with propagation of the waves, but which constantly changes from one streamline to another as a result of collisions with the hidden particles of the subquantum medium.

It seems to me that the random element must be introduced in order to complete the Theory of the Double Solution. The particle is then in constant contact with a "hidden thermostat" and its motion undergoes continuous fluctuations about the regular motion predicted by the guidance formula. As a result of this there exists the possibility of postulating a special system of "thermodynamics" for a single particle, because the particle is never really isolated. This idea has already been developed satisfactorily [13] and it may come to play an important role in the future progress of quantum physics.

3 – *Einstein's Views on Waves and Particles.*
Introduction of Non-linearity into the Theory of the Double Solution

My first attempts to interpret wave mechanics in terms of the Theory of the Double Solution in 1926–1927, were undoubtedly suggested to me by Einstein's work on general relativity. Einstein believed that the physical world should be described wholly by means of fields, well-defined at every point of space–time and obeying well-defined equations of a non-random nature. His theory was, essentially, that the entire physical reality, including particles, could be described by the appropriate solutions of field equations.

In the idealised theory which he visualised, there should be no terms in the equations which would represent field sources (like the charge and electric current terms in the second part of the Maxwell–Lorentz equations). If field source terms were not excluded, the differential field equations would be inadequate to determine the evolution of the total field, even if the initial and boundary conditions were given.

Einstein's attitude in no way indicated that he denied the existence of particles. On the contrary, he considered their existence to be an indisputable fact, but he regarded the particle not as an additional unit external to the field, but as an entity incorporated in the structure of the field and constituting a kind of local inhomogeneity. Einstein considered that all fields existing in nature, whether gravitational, electromagnetic or other fields, are perhaps only different manifestations of a single fundamental field, and should always contain very small regions of extremely high value, corresponding to the usual picture of particles which thus become *incorporated* in the field. The expression "bunched fields" has been used to describe this type of field.

Transposed into wave mechanics, Einstein's general theory, which involved the incorporation of the particle in the field, led naturally to the wave defined in the Theory of the Double Solution, since the u-wave, is in fact a bunched field. However, here we depart somewhat from Einstein's concepts by requiring that the field should be a wave field, so that the fundamental formulae $W = h\nu$ and $\lambda = h/p$ can be retained. Thus, as in general relativity, the field equations should be non-linear and the motion of the bunch, i.e. of the particle, should result from this non-linearity.

Einstein drew particular attention to the following very important point. If the equations for a certain field are linear, a singular solution of these equations can always be found in which the motion of the singularity is arbitrarily chosen in advance. A regular solution can always be added to this singular solution without affecting the motion of the singularity. We can infer that if the field equations are linear, there can be no guidance for a region with high field values by the regular solution. This is not so if the equations are non-linear, since we no longer obtain a solution by adding together several solutions, and the motion of the singularity in a singular solution may turn out to be determined by the external regular form of the solution. We shall return soon to this very important point.

Thus, it now appears that the guidance of a singularity by a continuous wave, such as is postulated by the Theory of the Double Solution, must imply non-linearity of the wave equation. The non-linearity which is thus introduced into the wave propagation must in all cases which are described precisely by the usual theory be *very localised*, that is to say, this non-linearity should only be significant in the very small region which constitutes the particle in the true sense of the word. It is only in this very small region, which in general is in motion and where the u-wave assumes a very large value, that the non-linear terms of the wave equation become important. Outside this region, the non-linear terms should be negligible and the equation for the u-wave should obviously reduce to the linear equation for the Ψ-wave. This is so because we know that the Ψ-wave enables us to predict observable physical phenomena such as the effects of interference and diffraction or even, by calculating the eigenvalues, the energies of the stationary states of an atom. This fact inevitably leads to the conclusion that the equation for the u-waves must agree *almost everywhere* with the linear equation of the Ψ-waves. It is only in the small singular regions (particles) where the values of u are very high (and perhaps on the sharp edges of certain wave trains where the derivatives of u can become very large) that non-linearity can be manifested.

We end with a comment which could be of great importance. Everything we have just said applies only to states which can be described in wave mechanics by means of Ψ-waves, since the equation of linear propagation of the Ψ-wave is known to give good results. This makes it necessary to consider a very localised non-linearity. However, there are microphysical processes which present-day quantum physics cannot describe, for example, the quantum transitions between stationary states of quantized systems, but it is probable that there are other processes, the description of which evades the present-day theory. In these cases, it may be that during the lifetime of transient processes of very short duration, the non-linear terms become very significant, even outside singular regions. These transient processes cannot be described by usual theory which is essentially linear. In other words, the usual theory can only describe steady-state states, when the non-linear terms cancel out everywhere except for the very localised singular regions. These ideas, which seem to be related to certain results of the theory of limit cycles in non-linear mechanics, could open up a field of research which, though difficult to explore, is of great interest, and which is especially capable of providing a lucid explanation of quantum transitions. Some very interesting papers by Fer, Lochak, Andrade e Silva and Leruste seem to confirm this hope [6].

4 – The Form of the u-Wave and the Relationship between the u-Wave and the Ψ-Wave

Despite its interest, we must now leave the case of transient processes and consider the nature of the u-wave. We shall assume that the objective u-wave conforms to an equation which is non-linear in the singular region at high values of the wave field, but which obviously reduces to a normal linear equation of wave mechanics (Schrödinger, Klein–Gordon, Dirac, etc.) outside this region. In the linear domain it should be possible to find a solution u_0 which is very small at large distances from the singular region but which increases very rapidly on approaching it and would contain a mathematical singularity in this region if the linear equation remained valid there. It has been possible to actually evaluate u_0 in certain special cases. It should also be possible to find a continuous solution v of the usual type of wave mechanics such that the solution u of the non-linear equation could be written as

$$u = u_0 + v \tag{8}$$

where $u_0 \ll v$ within the entire linear domain outside the singular region.

This solution will continue into the non-linear singular region, but the resolution into u_0 and v will then no longer have any meaning. The non-linearity ruling in this region will have the effect that u_0 and v will not be independent. The relationship between them is given by the guidance formula and consists in the fact that the trajectory of the u_0-singularity must coincide with one of the streamlines of the v-wave. This means that in this region the u-wave always remains in phase with the external encompassing wave. By associating the singular region with the particle, we return to the picture which I had in my early researches: the very small singular region constituting the particle is the seat of a periodic phenomenon which can be likened to a clock and which moves in the midst of the v-wave, of which it is an intimately integral part, in such a way as to remain constantly in phase with it.

We shall now specify precisely the relationship which must exist between the u-wave and the Ψ-wave. To do this, we must distinguish between the v-wave, i.e. the regular part of u with an objective character, and the Ψ-wave of normal wave mechanics—the simple subjective representation of probabilities—while considering them as being linked intimately with the normal linear wave equation.

This distinction, which is on the same basis as the Theory of the Double Solution and justifies its name, will enable us to eliminate the simultaneous objectivity and subjectivity which we are forced to attribute to the Ψ-wave in the usual interpretation of wave mechanics.

Actually, since the Ψ-wave, which is a representation of probability, must be constructed according to our information on the state of the particle, we can define it, if our information is precise, as being everywhere proportional to the v-wave so that

$$\Psi = Cv, \qquad (9)$$

where C is a normalisation constant*. Since the u-wave, and consequently its external part v, are assumed to have an objective reality, they should have everywhere a well-defined value. On the contrary, the Ψ-function can be normalised at will by a suitable choice of the constant C. The Ψ-wave defined by equation (9) satisfactorily conforms everywhere to the normal linear equation. Moreover, the Ψ-wave would not conform to a non-linear equation because the principle of superposition appears to be a necessary condition of the normal statistical interpretation, as Pauli had previously emphasised in his article on wave mechanics in *Handbuch der Physik*.

The mystery of the simultaneously objective and subjective nature of the wave in the customary interpretation thus appears to have vanished. The objective v-wave can determine physical phenomena such as interference, diffraction, and quantized energy states of atomic systems; the Ψ-wave is only a representation of probability and is subjective in character and can be normalised at will. Since, however, the Ψ-wave must imitate the v-wave in accordance with (9), the impression has been given that it is the Ψ-wave which gives rise to the physical phenomena mentioned above. This explains why we have been led to

---

* As Destouches has remarked, it would be more precise, instead of putting $\Psi(x, y, z, t) = Cv(x, y, z, t)$, to write

$$\Psi(x_1, y_1, z_1, t) = C[v(x, y, z, t)]_{x=x_1, y=y_1, z=z_1},$$

where x_1, y_1, and z_1 are the coordinates of the particle.

attribute to the Ψ-wave a hybrid character which is very unsatisfactory. I believe that the interpretation which we have just given within the framework of the Theory of the Double Solution could quite well be the only one capable of resolving the riddle of the subjective *and* objective character of the wave normally envisaged in wave mechanics.

Moreover, it seems quite likely that this approach will also resolve the problem of the "reduction of the wave packet" without making a physically incomprehensible phenomenon of it. Whilst a physical process (for example, the effect of a measuring device) dissociates the u-wave of the particle into separate portions, with breakdown of phase relationships, and after verification we learn that the particle is present in one of these portions, we must, in order to describe the new state of our knowledge, put $C = 0$ for all regions other than that where we now know the particle is found, and re-normalise the Ψ-wave in this region. Thus, the possibility of having different, including zero, values for the constant C for the different components into which the u-wave is split up, enables us to interpret the reduction of the wave packet without affecting the objective character of the u-wave. This is perhaps the only reasonable physical interpretation of the reduction of the wave packet.

Let us now state more precisely a little of what we have just said. In the initial state the wave function is $\Psi = Cv$. After breakdown of the phase relationships between the portions v_1, v_2, ... of the wave by the process of measurement, we must put $\Psi = C(v_1 + v_2 + \ldots)$. However, having acknowledged the presence of the particle in one of the portions of the wave, say v_k, we must put $C = 0$ for all the v_i except v_k and $C = C'$ for v_k, so that now we shall be able to define the subjective probability wave by $\Psi = C'v_k$, where C' is a new normalisation constant *.

5 – *A new Way of Considering the Guidance Formula*

In the general theory of relativity, authors such as Georges Darmois, Einstein and his colleagues, and more recently André Lichnérowicz, Pham Tan Hoang, and others, have established that a moving material particle must follow a geodesic of the external field by virtue of the

---

* We note again that the presence of the normalisation factor C in equation (9) allows the Ψ-wave to conserve the properties of propagation of a physical wave, but deprives it of the property of amplitude addition which is characteristic of physical waves. This explains the hybrid nature of the Ψ-wave.

non-linear equations of the theory. They have, in fact, shown that a field singularity defining a material particle must, in space–time, remain inside an extremely thin tube, the walls of which are formed by geodesics of the external field. In other words, the very high values of the field which constitute the particle, must remain "imprisoned" inside this tube. Any other form of the world tube representing the motion of the particle thus defined would not be compatible with the existence of a continuous solution of non-linear equations for the gravitational field.

In a remarkable little book [7], Georges Darmois presented this theory in a particularly striking fashion. He wrote: "The theory proposed by Einstein to replace that of Newton, instead of linking the masses by force, links them by the field into which they are integrated: the world tubes which describe the motion of the material masses are "immersed", so to speak, in the same field, and it is this which creates their interdependence." Later on he added: "It is in order to interlock its own field with the external field that a small mass must follow a geodesic."

He had also made some sound comments, which have since been developed by Lichnérowicz [8], on the manner in which an attempt can be made to extend the external field towards the inside of the world tube of the particle. He regarded as obvious the fact that this external, extensive field must contain a singularity and added with much profundity: "This fundamental role of singularities, which in a way foreshadow mass tubes, is extremely important."

Although this problem is quite different, the question of guidance of the particle in the Theory of the Double Solution can be approached in a similar manner. Outside the world tube of the particle, the u-wave reduces itself to the v-wave, which can be considered as constituting the external field. But the solution $u = u_0 + v$ of the external wave equation, if it could be extended uniformly to the interior of the tube, would represent a singularity there. The similarity with the ideas of Georges Darmois is striking, and the same reasons for continuity which had guided him, lead here to the following conclusion: the walls of the very thin world tube which contain the high values of the field, i.e. the particle, are formed by stream-lines of the external field; the particle is thus imprisoned in the world tube, which influences the guidance formula. Paraphrasing a quotation made above by Darmois, we can say: "It is, on the whole, in order to interlock its internal wave field with the external wave field v, that the particle must describe a streamline."

CHAPTER V

Critical Study of Certain Points of the Usual Interpretation of Wave Mechanics

1 – *Statement of the Theory of Transformations*

The usual interpretation of wave mechanics rests partly on a formalism which is often called the Theory of Transformations. Our task in this section will be to subject this formalism to a critical review.

Our starting point is the observation that in wave mechanics one regards every physical quantity as corresponding to a linear and Hermitian operator, A, which has a series of complex eigenfunctions φ_i. These eigenfunctions form a complete orthogonal set so that a wave function Ψ can always be expanded in the form

$$\Psi = \sum_i c_i \varphi_i \tag{1}$$

where c_i are the complex generalised Fourier coefficients. They are given by

$$c_i = \int \varphi_i^* \Psi \mathrm{d}\tau. \tag{2}$$

[If the spectrum of c_i is continuous, then equation (1) still applies, but the φ_i are now the "eigen-differentials" of the spectrum. These eigen-differentials represent groups of waves whose introduction can be justified physically by noting that monochromatic waves are abstractions and that only groups of waves have a physical reality.]

The quantities c_i are the "coordinates" of Ψ in the Hilbert space relative to a system of reference defined by the basic functions φ_i. If the coefficients c_i are known, then Ψ is known. If we transfer from the set of basic functions φ_i to another set of basic functions φ_i' then, since the entire theory is linear, we have

$$\varphi_i = \sum_k d_{ki} \varphi_k' \tag{3}$$

and consequently

$$\Psi = \sum_i c_i \varphi_i = \sum_k c'_k \varphi'_k \qquad (4)$$

where

$$c'_k = \sum_i d_{ki} c_i. \qquad (5)$$

Moreover, it is assumed that the eigenfunctions φ_i which correspond to the position vector $\mathbf{R_0}$ of the particle are the Dirac functions $\delta(\mathbf{R}-\mathbf{R_0})$. We have already seen that this hypothesis is physically dubious because observable particle localisation results, in fact, from a short-range effect between the particle and another microphysical system, which gives rise to an observable phenomenon by a chain process, and not by a reduction of the wave function to a Dirac δ-function. If we accept, as is customary, the doubtful postulate in question, we can write

$$\Psi(\mathbf{R}) = \int \Psi(\mathbf{R_0}) \delta(\mathbf{R}-\mathbf{R_0}) \, d\mathbf{R} \qquad (6)$$

and we can consider $\Psi(\mathbf{R_0})$ as being the coefficient c_i in expansion of Ψ in terms of the eigenfunctions of position.

Moreover, we note that if the φ_i are functions of x, y and z, and the coefficients c_i can be time-dependent, then where (1) is substituted into the wave equation

$$\frac{h}{2\pi i} \frac{\partial \Psi}{\partial t} = H\Psi, \qquad (7)$$

where H is the Hamiltonian operator, we have, after multiplying by φ_i^* and integrating with respect to $d\tau$,

$$\frac{h}{2\pi i} \frac{\partial c_j}{\partial t} = \sum_i H_{ji} c_i, \qquad (8)$$

where $H_{ji} = \int \varphi_j^* H \varphi_i \, d\tau$ is the matrix element corresponding to the operator H in the φ_j system. Equation (8), which is Dirac's equation for the variation of constants, can be considered as equivalent to the wave equation.

This elegant formalism suggests that all "representations" of the Ψ-function by complete and orthonormal systems of eigenfunctions corresponding to physical quantities are equivalent from a physical point of view. In particular, the "q-representation", i.e. the coordinates representation involving equation (6), should be just as significant as the "p-representation", i.e. the momentum representation which is defined in terms of the expansion coefficients $c(\mathbf{p})$ relating plane monochromatic waves. The wave equation (7) should have no advantage over the equations (8) for the variation of constants.

It is a general postulate of the Theory of Transformations that "the probability that a quantity A associated with a particle will have as its value the eigenvalue α_i of the corresponding operator is equal to $|c_i|^2$". When the general law is applied to the q-representation, it is found that the probability of the presence of the particle at the point $\mathbf{R_0}$ is $|\Psi(\mathbf{R_0})|^2$, which is the principle of localisation.

Before we examine the significance of this deduction, we must make an important comment. We have frequently stated the general law of probability in the form: "The probability that a quantity A has the value . . ." We ought really to say: "The probability that a measurement (or better still, an observable phenomenon) enables us to assign to the quantity A a value equal to . . ." I think that even the strongest supporters of the present-day interpretation would agree that this second statement alone is correct. Now, the first statement differs essentially from the second, for it accepts implicitly that the quantity A has an *a priori* probability $|c_i|^2$ of having the value α_i prior to any measurement. This does not appear to be justified in any way, and does not agree with the idea that the process of measurement modifies the state which exists prior to its intervention. This is an important comment, which may imply that erroneous conclusions have been drawn from the Theory of Transformations.

2 – *Critique of the Theory of Transformations*

The Theory of Transformations is open to criticism even from a purely formal point of view. It is not true to say that there is a complete equivalence between the set $c_i \varphi_i$ *considered individually* and the sum $\Psi = \sum_i c_i \varphi_i$. In fact, the coefficients are complex numbers of the form $c_i = |c_i| e^{i\beta_i}$, and if the $c_i \varphi_i$ are considered separately, the "phase differences" $\beta_i - \beta_j$

are not made apparent, whilst in the sum $\sum_i c_i \varphi_i$ these phase differences are very important. The sum $\Psi = \sum_i c_i \varphi_i$ contains interference terms which are not present at all when the set $c_i \varphi_i$ is considered individually. In particular $|\Psi|^2$ cannot be derived from the set $|c_i|^2$.

Furthermore, I think it is incorrect to make the eigenfunction $\delta(\mathbf{R}-\mathbf{R}_j)$ correspond to particle localisation at the point $\mathbf{R}_j$ since, as we have already said, observable localisation does not arise from reduction of the wave to a point, but from a short-range effect between two microparticles. In particular, the passage of the wave through a small hole in a screen, which is not accompanied by any observable effect, does not in any way constitute a particle localisation. Particle localisation is never observed in this way.

However, the deceptive nature of the Theory of Transformations is much more obvious from a physical point of view. In order to explain it, we must remember that the entire theory originates from classical physics, or more precisely, from d'Alembert's theory of vibrating strings. Let us examine this situation in greater detail.

We shall consider a uniform string which is stretched along the x-axis with its ends fixed at the points $x = 0$ and $x = l$. When the string is displaced in the y direction and then released, it will vibrate in the xOy plane. The vibration may be represented by the superposition of complex functions of the form

$$\varphi_n = c_n \sin n\pi \frac{x}{l} e^{2\pi i \nu_n t} = c_n \sin 2\pi \frac{x}{\lambda_n} e^{2\pi i \nu_n t} \tag{9}$$

which are the eigenfunctions of the problem with $\lambda_n = 2l/n$ and $\nu_n = V/\lambda_n = nV/2l$, where V is the velocity of the waves along the string. The motion of the string will be given by

$$y = \sum_n c_n \varphi_n = \sum_n \left(c_n \sin n\pi \frac{x}{l} e^{2\pi i \nu_n t} + c_n^* \sin n\pi \frac{x}{l} e^{-2\pi i \nu_n t} \right). \tag{10}$$

If this motion is observed by a photographic method, it is found that the string assumes in general a very complicated shape which varies with time. However, we shall not be able to observe the harmonics separately unless we disturb the motion of the string. Certainly, if the function $y(x, t)$ is known, then the harmonics can be calculated theoretically and then superimposed to form $y(x, t)$. However, the analysis of y into a harmonic series exists only in the mind of the theoretician—it does not exist in the visible motion. Nevertheless, these harmonics could be separated physically, but it would then be necessary to use devices which, in separating them, will modify the phase relationships which exist between them during the motion of the string. For example, if the string is placed in a gas such as air, it will emit a sound which will correspond to the function $y(x, t)$ and will consequently be formed by a combination of harmonics. The corresponding frequencies can be determined by arranging a number of Helmholtz resonators with resonant frequencies equal to the various ν_n around the string. The various harmonics of the sound emitted can thus be separated, but their phase relationships will have been modified so that the phase differences which determined the shape of the string will not be known.

A similar result is obtained by considering a radio antenna carrying a current formed by the superposition of harmonics of a fundamental frequency. The electromagnetic wave emitted will be a superposition of harmonics and will have a very complex form. However, the oscillatory circuits tuned to the various harmonics and placed in the wave field will become the seat of harmonic currents. The various harmonics of the incident wave will thus have been separated, but our knowledge of the currents in the oscillatory circuits will not allow us to reconstruct the exact waveform because we shall have lost the knowledge of the phase differences.

An example which perhaps closer approaches wave mechanics is that of a wave train (for example sound or elastic waves) formed by the superposition of plane monochromatic waves incident on a device such as a prism or a grating. The incident wave, if it can be recorded, corresponds to a complex motion of the medium through which it passes, and even if a knowledge of this motion enables a theoretician to calculate the monochromatic components of the superposition, these components do not exist in a separated fashion in the incident wave, but only in the mind of the theoretician. The effect of a device such as a prism or a grating is to separate the monochromatic components and to concentrate them in different directions in space. The wave is thus finally subdivided into essentially monochromatic waves which are spatially separated. Thus, whilst we shall have separated the monochromatic components, the phase relationships which existed between them and which determined the incident wave form will no longer be manifested because of the spatial separation.

A consideration of the foregoing examples, and others which we could imagine, necessarily leads us to the conclusion that it is the representation in space and time which is objective, and not the Fourier analysis which only exists in the mind of the theoretician. The various Fourier components can only be observed by means of devices which change completely the initial state of affairs and modify the phase relationships.

In the language of the Theory of Transformations, this can be expressed by saying that the q-representation is the only objective representation, whilst the p-representation—the abstract representation in momentum space—exists only in the mind of the theoretician. This shows, contrary to what the Theory of Transformations usually asserts, that the two representations—q and p—are by no means equivalent. It is the wave function that describes the physical reality and not the coefficients c_i considered separately. Moreover, this conclusion is the consequence of the obvious fact that three-dimensional space is a physical reality and is the essential framework of our experiment, whilst momentum space is only an abstract mathematical representation.

3 – Fundamental Importance of the q-Representation. Actual and Predicted Probabilities

The above analysis suggests that the probability of location ρ, which is equal to $|\Psi|^2$ in the case of Schrödinger's equation, has a kind of supremacy over the other probabilities visualised by the usual theory, because it corresponds to the presence of the particle at a point on the wave before a device which separates the Fourier components with breakdown of the phase relationships is employed. For quantities other than the location probability and those derived from this probability (that is to say, in abstract language, for quantities which do not commute with position), the probabilities $|c_i|^2$ correspond to the situation which exists *after* the action of a measuring device. This device separates, with breakdown of phase relationships, the components φ_i which are associated with the possible eigenvalues α_i of the quantity A when the result of the measurement is still unkown. The effect of the measuring device must be to detach the particle from its initial wave in order to re-attach it to one of the spectral components. In the language of the Theory of the Double

Solution, this means that during disturbance of the wave which is brought about by the action of the measuring device, the particle is *guided* in such a manner that this result is finally obtained. The attachment of the particle to one of the spectral components (with loss of the phase relationships) can take place either by spatial separation of the spectral components (in the case of devices of the prism or grating type), or by a process of "shunting" the particle into one of the spectral components. I have studied these questions in my book *Théorie de la Mesure* [2], but they are outside the scope of this book.

Briefly, the probability density ρ appears to me to exist in the initial state prior to the action of any measuring device, whilst the probabilities $|c_i|^2$ only came into play after the action of the device which measures the quantity A, with the eigenfunctions φ_i. The $|c_i|^2$ cannot have the significance of probabilities existing objectively in the initial state. This circumstance appears to me to make it impossible to regard the $|c_i|^2$ as representing the probabilities existing simultaneously in the initial state. The final proof of this lies in the fact that in the initial state, the measurement of any quantity whatever is possible *a priori*, and that the nature of the measuring process will, in general, influence the $|c_i|^2$. I have already mentioned the very curious fact that a theory which relies on the concept that every process involving a measurement necessarily disturbs the state of the system, should place ρ and $|c_i|^2$ on an equal footing and thus ignore its own fundamental assumption.

In order to explain this question more precisely, it is very important to emphasise the essential difference between a *real* probability, valid at the instant when it is estimated (as we assume to be the case with the probability ρ), and the probabilities which are simply *predicted* to correspond to future situations, e.g. the $|c_i|^2$ for the initial Ψ-state.

A simple example of this can be given. Suppose that I have before me a rotating roulette wheel with an equal number of red and black compartments, and that I have a ball in my hand. In this initial situation, the probabilities of presence of the ball are

in my hand: 1,
in a red compartment: 0,
in a black compartment: 0.

These probabilities correspond to the actual situation. But I can predict the probabilities which will be valid when I have thrown the ball on to the roulette wheel and when the ball ultimately stops in one compartment, the colour of which will be, as yet, undetermined. These predicted probabilities will be

(A) in my hand: 0,
　　in a red compartment: $\frac{1}{2}$,
　　in a black compartment: $\frac{1}{2}$.

But there are no valid simple *predicted* probabilities so long as I still hold the ball in my hand. The "not yet valid" nature of these probabilities can be demonstrated by the fact that if I had by my side another roulette wheel, all the compartments of which were black, then the predicted probabilities would be

(B) in my hand: 0,
　　in a red compartment: 0,
　　in a black compartment: 1.

These predicted probabilities are incompatible with the foregoing and I should not know whether it is the probabilities (A) or (B) which would become real when I decide to throw my ball on to one or other roulette wheel.

These considerations can be directly applied to the probabilities of wave mechanics. In an initial state corresponding to a known wave function Ψ, the probability of localisation $|\Psi|^2$ is real. For every measurable physical quantity there are *predicted* probabilities $|c_i|^2$, which will only become valid after the intervention of the process of measurement. As long as no definite quantity is measured, it will not be known which of the set of $|c_i|^2$ will actually become real probabilities. It seems important to consider these questions in order to avoid confusion and improve our understanding of the importance which we propose to attribute to the probability of location.

4 – *Significance of the Uncertainty Relationships*

The formalism of the Theory of Transformations has led the present-day interpretation of wave mechanics to consequences which now appear to be probably inaccurate.

We shall first consider the uncertainty relationships between two canonically conjugate quantities, for example, $\Delta x \, \Delta p_x \geqq h$ (or in a more precise form $\sigma_x \sigma_{p_x} \geq h/4\pi$). What is the significance of the uncertainties Δx and Δp_x? In the case of Δx the answer appears to be obvious: it is the uncertainty on the coordinate x of the particle which can be found at any point whatever of the wave with the probability $|\Psi|^2$. The only disputable point is the significance of the words "can be found". Do they mean that the particle has a position in the wave train but this position is unknown to us, or do they mean that the particle has no position in the wave train, but it is potentially present in the entire wave train? We know that it is the second hypothesis (much less clear than the first) which is adopted by the present-day interpretation. We must now by-pass this point and ask: what is the significance of Δp_x? According to the modern interpretation, Δp_x is the uncertainty in p_x in the state Ψ, the potential values of p_x being distributed within the whole of the interval Δp_x. Thus, the uncertainties Δx and Δp_x would be "real" in the state characterised by the function Ψ.

We do not consider this interpretation to be acceptable. Without doubt, Δp_x appears to be the *predicted* uncertainty in the value which p_x can assume *after* the action of a device which is used to measure p_x, the prediction being made before the result of this measurement becomes known. Moreover, since it is a leading principle of the usual interpretation that the action of the measuring device completely disturbs the initial state, the uncertainty Δp_x cannot relate to the same state as Δx. As far as we are concerned, the uncertainty Δx really exists in the initial state prior to any measurement of the momentum (when the Fourier components of the wave still interfere with each other), whilst the uncertainty Δp_x is then only a predicted uncertainty which becomes real only after the action of the device for measuring the momentum, when the Fourier components of the wave have ceased to interfere, and when the particle remains attached to one of them. The uncertainty relationship $\Delta x \Delta p_x \geq h$ thus has quite a different meaning from that generally given to it, since the uncertainties Δx and Δp_x are only real in *different* states of the particle. Moreover, there is no compulsion to consider these uncertainties as the true uncertainties in x and p_x as the customary theory does. It is much more natural to consider them as simple uncertainties due to our lack of knowledge of the true values of x and of p_x. In this way we return to the traditional idea of probability which is introduced because of our ignorance of something which exists objectively.

In the Theory of the Double Solution, the particle occupies at every instant a position in the wave which is unknown but which exists, and hence the uncertainty Δx. If we knew the position of the particle, the guidance formula would enable us to calculate its momentum, and hence the value of p_x. However, since we do not know this position, there is a resulting uncertainty in the value of p_x in the initial state; the value exists but we do not know what it is. But this real uncertainty in the initial state is not the uncertainty Δp_x of the Heisenberg relation, because it is only in the initial state that this is a predicted uncertainty: it does not become actual until after the measurement of p_x which completely modifies the structure of the wave.

The same kind of analysis would be carried out for any uncertainty relation of the form $\Delta A \Delta B \geq a$, where both ΔA and ΔB are predicted uncertainties relating to two different situations which cannot exist after the completion of the two incompatible measurements.

5 – The Statistical Structure of Wave Mechanics

These concepts are quite differnet from those of the usual interpretation and lead to a revision of the entire statistical structure accepted in this interpretation. This statistical structure is, in fact, completely at variance with the classical calculus of probabilities. I studied this question in an article in *Revue Scientifique* in 1948 [9] at a time when I still accepted the orthodox interpretation, but not without a certain uneasiness. This undoubtedly contributed to my turning towards another interpretation.

In the calculus of probabilities, a probability density $\rho_X(x)$ is defined for a continuous random variable X, so that $\rho_X(x)\mathrm{d}x$ is the probability that X has a value included between x and $x+\mathrm{d}x$. For another random variable Y, a probability density $\rho_Y(y)$ is similarly defined. Next, a probability density $\rho(x, y)$ is defined so that $\rho(x, y)\mathrm{d}x\mathrm{d}y$ is the probability of obtaining in the same measuring operation (the same "trial" as the statisticians say) values of X and Y included within the intervals $x \to x+\mathrm{d}x$ and $y \to y+\mathrm{d}y$ respectively. Finally, one can introduce the probability density $\rho_Y^{(X)}(x, y)$, which corresponds to the probability of obtaining the value y for Y *when it is known* that X has the value x. In the same way one can define analogous probability $\rho_X^{(Y)}$. It is easy to see that the five probability densities which we have just defined, must satisfy the following relationships:

$$\rho_X(x) = \int \rho(x, y)\,dy, \qquad \rho_Y(y) = \int \rho(x, y)\,dx,$$

$$\rho_X^{(Y)}(x, y) = \frac{\rho(x, y)}{\rho_Y(y)}, \qquad \rho_Y^{(X)}(x, y) = \frac{\rho(x, y)}{\rho_X(x)};$$

(11)

from which we conclude that

$$\rho_X(x) = \int \rho_X^{(Y)}(x, y)\,\rho_Y(y)\,dy,$$

$$\rho_Y(y) = \int \rho_Y^{(X)}(x, y)\,\rho_X(x)\,dx.$$

(12)

The whole of this statistical structure results very clearly from the concrete probability picture when we imagine the "individuals" for which the quantities X and Y have determined values, statistics being introduced by the simultaneous consideration of individuals for which X and Y have different values.

However, this statistical structure is not applicable to the probabilities defined in the usual interpretation of wave mechanics. On the one hand, the probability $\rho(x, y)$ does not generally exist here, and on the other hand, the products $\rho_X^{(Y)}(x, y)\,\rho_Y(y)$ and $\rho_Y^{(X)}(x, y)\,\rho_X(x)$, which should be equal, are in fact not equal. This failure of the classical statistical structure in the usual interpretation is explained, in my opinion, by the fact that the probabilities which are involved in this interpreation are not related to the same state of the particle, and that they are not simultaneously "real". This leads to a failure even of classical statistics, which are not applicable to these probabilities. However, if in accordance with the Theory of the Double Solution, we assign to a particle occupying a certain position the momentum defined by the guidance formula, then we can re-establish the validity of classical statistics both in the initial state, prior to the action on the wave of the momentum device, and in the final state which follows the action of this device. I have shown this in detail especially on page 88 et seq. of my book on *Théorie de la Mesure* [2].

This leads us to refer to the well known theorem of von Neumann, according to which it is impossible to give an interpretation of the laws of probabilities of wave mechanics by a picture which introduces hidden variables and which thus permits a sharply defined localisation and momentum to be assigned to the particle.

Von Neumann demonstrated this theorem thirty years ago, starting from the very elegant formalism of statistical matrices. He proceeded to show, apparently very rigorously, that the wave mechanical probability distributions cannot be reduced to a statistical system of the classical type by the introduction of hidden variables.

Now, the very fact that we can, as we have said, restore the classical statistical structure by using the hidden variables of the Theory of the Double Solution, shows that von Neumann's Theorem cannot be exact, even if the picture proposed in the Theory of the Double Solution does not conform with physical reality. It is sufficient to cite a single counter-example, even without physical reality, to prove the fallacy of the prohibition which seems to result from von Neumann's reasoning.

In our opinion, the error in this reasoning is as follows: it lies essentially in the postulate that in the state represented by Ψ, the probability distributions for two canonically conjugate variables are both simultaneously *real*. However, although this postulate was suggested by the Theory of Transformations and is currently accepted, it now appears to us to be completely inaccurate for the reasons explained above. For example, by neglecting the unavoidable effect which every process of measurement of the momentum exercises on the initial state, it disregards the fact that the location probability $|\Psi|^2$ and the momentum probability $|c_p|^2$ cannot simultaneously be real in the initial state. This fact, which takes into account the unavoidable effect of the process of measurement and appears to be indisputable to the supporters of the usual interpretation, causes von Neumann's theorem to collapse, thus making it a pseudo-theorem.

6 – *Impossibility of Obtaining Interference Fringes and of Determining, at the Same Time, the Trajectory of the Particle*

We have seen that it is incorrect to state the concept of complementarity by saying that the particle never appears simultaneously in its granular and wave aspects. On the contrary, the record of interference or diffraction fringes obtained on a photographic plate results from an infinite number of tiny local spots which display the successive arrival of particles, whilst the set of fringes is a statistical effect of the wave aspect.

However, where the supporters of complementarity and of the potential presence of particles in the entire region of space appear to triumph, is when they show that in the case of interference the trajectory followed by the particle cannot be determined in any way whatever. Thus, in Young's classical double-hole experiment, if the device is mounted in such a manner as to produce fringes, it cannot be stated through which hole the particle has passed, and we conclude that either the particle has passed through neither of the two holes, or it has passed "potentially" through both holes. We must examine this problem.

To start the study of Young's double-hole arrangement, let us reproduce the argument put forward by Niels Bohr.

Suppose that the monochromatic wave train which is incident on the front face of Young's screen originates from a slit in another screen so that the latter plays the role of a light source. We shall designate the distance between the holes by a, the distance from the two screens by $D \gg a$, and the wave length of the particles by λ. Finally, we shall choose the axes x and y as shown in Fig. 6.

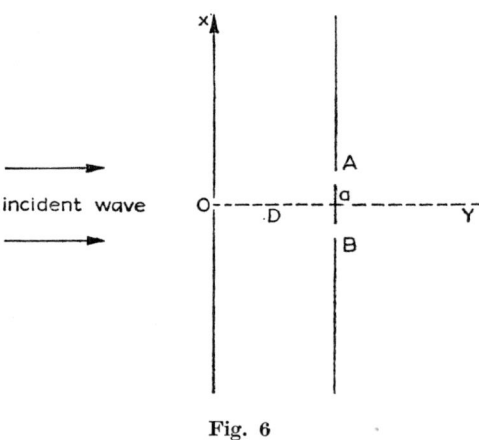

Fig. 6

The y-axis is equidistant from the two holes, and the primary slit is located on this axis. But, says Bohr, its exact position has an uncertainty Δx. Since in practice a and Δx are always very small in comparison with D, the phase difference between the waves which are incident on the two Young holes will be

$$\Delta \varphi = \frac{2\pi}{\lambda} \left[\sqrt{D^2 + \left(\frac{a}{2} + \Delta x\right)^2} - \sqrt{D^2 + \left(\frac{a}{2} - \Delta x\right)^2} \right] \simeq 2\pi \frac{a \Delta x}{\lambda D}. \quad (13)$$

In order that we may have sharp fringes it is necessary that the phase difference between the waves proceeding from the two holes be well defined, i.e. the uncertainty in the phase difference should be very much less than 2π. This condition gives us

$$\Delta x \ll \frac{\lambda D}{a}. \tag{14}$$

On the other hand, in order to be able to say through which of the two holes the particle has actually passed, after traversing the primary slit, it would be necessary to know the momentum of the particle on leaving the primary slit to a reasonable accuracy. If p_X and p_Y are the components of this momentum, the point on the second screen reached by the particle will have the abcissa x equal to Dp_X/p_Y and, if p_X is assigned an uncertainty Δp_X, this abcissa is assigned an uncertainty $D\Delta p_X/p_Y$. In order to be certain that the particle will pass through one of the two holes, it is now necessary to have

$$a \gg D\frac{\Delta p_X}{p_Y}. \tag{15}$$

However, the bundle of waves issuing from the primary slit is nearly parallel to the y-axis so that we have approximately $p_Y = p = h/\lambda$ and as a result, the foregoing condition can be written as

$$a \gg D\Delta p_X \frac{\lambda}{h}. \tag{16}$$

Now, we know that whatever the devices used for measuring the coordinates of the particle and the components of its momentum, we always have the Heisenberg inequalities:

$$\Delta x \Delta p_X \geqq h. \tag{17}$$

Condition (16) thus gives, *a fortiori*:

$$\Delta x \gg \frac{\lambda D}{a}. \tag{18}$$

Since conditions (14) and (18) are obviously contradictory, it can only be concluded that if we can say precisely through which of the holes the particle has passed, then it is impossible to observe interference phenomena, and conversely, if interference phenomena can be observed, we cannot say through which hole the particle has passed.

Naturally, the foregoing reasoning is also equally applicable to an electron, or another particle, as to a photon.

At first sight, Bohr's reasoning appears to be very elegant but on closer examination it is found to be open to criticism. The manner in which the uncertainty relationship (17) is introduced here is somewhat peculiar in that it assumes implicitly that the component p_X of the momentum of the particle can be measured by the recoil of the first screen along the x-axis. This is impossible because this screen has a macroscopic mass and can be rigidly fixed *. Moreover, Bohr does not introduce the size of the slit in the first screen (this must not be confused with the uncertainty Δx), which plays an essential role in the phenomenon of diffraction and which enables the wave, after passing through the slit in the first screen, to reach the two Young holes. Finally, this reasoning, like many other arguments of the same nature, introduces the concept of a

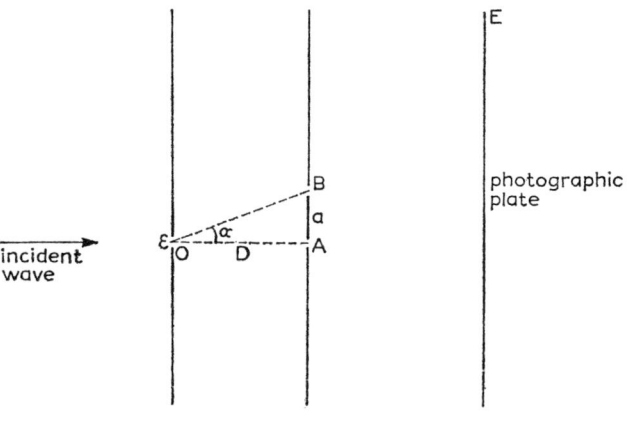

Fig. 7

---

* In this reasoning, the inequality (17) is introduced as if the first screen were comparable with the particle and could be set in motion by the exchange of momentum with the photon. This appears to me to be very questionable.

rectilinear trajectory for the particle between O and A or B, which is strange in a theory which declines to assign a trajectory to the particle.

It thus appears to be useful to take up Bohr's reasoning in a form which is nearer to the normal concepts of physical optics, in order to elucidate the true meaning of the above result. We commence by re-arranging Fig. 6 in a somewhat different way.

Here, ε denotes the size of the slit in the first screen. It is assumed that $a \ll D$ and also $\varepsilon \ll a$. In order to simplify the calculations, we suppose also that the lower edge of the slit in the first screen is just in line with the hole A.

The phase difference between the waves reaching the Young holes from the lower edge O of the slit is thus

$$\Delta\varphi = \frac{2\pi}{\lambda}(\sqrt{D^2+a^2}-D) \simeq \frac{\pi a^2}{\lambda D}. \tag{19}$$

On moving from the lower edge to the upper edge of the slit, the phase difference becomes

$$\Delta\varphi + \mathrm{d}\Delta\varphi = \frac{2\pi}{\lambda}(\sqrt{D^2+(a-\varepsilon)^2}-\sqrt{D^2+\varepsilon^2}) \simeq \frac{\pi a^2}{\lambda D} - 2\pi\frac{\varepsilon a}{\lambda D}, \tag{20}$$

so that

$$\mathrm{d}\Delta\varphi = -2\pi\frac{\varepsilon a}{\lambda D}. \tag{21}$$

In order to have interference fringes, it is necessary that $|\mathrm{d}\Delta\varphi| \ll 2\pi$, or,

$$\frac{\varepsilon a}{\lambda D} \ll 1. \tag{22}$$

However, since $\tan \alpha \simeq \alpha = a/D$, the theory of diffraction tells us that if $a/D \ll \lambda/\varepsilon$ then the wave will be highly diffracted at the slit in the first screen. Moreover, the amplitudes of the diffracted waves at A and by B will be approximately equal, and if condition (22) is satisfied then the phases of the waves transmitted through all points of the slits A and B will be approximately the same. Condition (22) is thus the condition which is necessary and sufficient for Young's fringes to be observable, but it does not tell us through which hole the particle has passed.

If, on the contrary, we have $a/D \gg \lambda/\varepsilon$, the wave will be weakly diffracted and it will not reach the hole B. Now, we can be certain that the particle which reaches the film and produces a photographic record has passed through hole A, but since all this takes place as if the hole B did not exist, there will no longer be interference, and Young's fringes will not be observed. The condition for knowing through which hole the particle has passed is thus

$$\frac{\varepsilon a}{\lambda D} \gg 1. \tag{23}$$

The inequalities (22) and (23) are identical with (14) and (18) which were obtained by Bohr's reasoning, but on condition that Δx is replaced by ε, which has a much clearer physical meaning. Since (22) and (23) are contradictory, it can be seen that it is impossible to observe interference *and* to know through which hole the particle has passed.

However, this conclusion, which confirms that of Bohr by a somewhat different method, by no means compels us to accept that the particle does not pass through either of the two holes, or that it passes at the same time through both holes in a potential manner(?). This can be understood from the standpoint of the Theory of the Double Solution. If condition (22) is satisfied, then the diffracted wave covers holes A and B uniformly and the stream lines of the wave pass in equal numbers through the two holes, which are assumed to be of equal area. The particle following *one* of these stream lines (Bohm–Vigier perturbations apart) will pass through one of the holes, but no observable particle localisation can be produced prior to its arrival at the photographic film, and it will not be known through which hole it has passed. If, on the contrary, inequality (23) is satisfied, then the slit in the first screen will be so large that the diffraction will be weak, and in practice the wave will not reach hole B. Now, all the stream lines of the wave which traverse Young's screen will pass through hole A and it will now be known through which hole the particle has passed, but obviously Young's fringes will no longer be obtained.

In order to compare our reasoning with that of Bohr, it may be noted that it involves the following essential points. There are two conditions for the appearance of Young's fringes, namely, we must know (a) the phase difference between the waves on holes A and B should be sharply defined, and (b) that the wave amplitudes on these two holes should be practically equal. Now, each one of these two conditions implies that size ε of the slit in the first screen should be small in comparison with the wavelength. The two conditions are met simultaneously when inequality (22) is satisfied. This is the first inequality in Bohr's reasoning (with the proviso that ε be substituted for Δx), but it occurs in the reasoning in order to satisfy condition (a), without reference to condition (b). When it is inequality (23) which is satisfied, then the slit in the first screen is sufficiently large for the phase difference between the waves at A and B to be no longer sharply defined, and there can no longer be interference there. But, *at the same time,* the diffraction is so weak that the wave does not reach the two holes A and B simultaneously: it will be known that the particle has passed through A, but there will no longer be interference fringes. Thus, by means of clear and classical wave concepts, a result is obtained which is similar to that of Bohr, but which appears to be very much closer to physical reality and no longer imposes a mixture of contradictory concepts (the non-existence of particle trajectories and the rectilinear trajectory of O to A or B).

One can visualise other ways of determining through which of the two Young holes the particle has passed. For example it could be supposed that close to one of the holes, let us say hole B, there is a microphysical system which reacts through a short-range effect to the passage of the particle through this hole and gives rise to an observable localisation by a chain process. The passage of the particle through hole B could thus be detected, but it is quite obvious that the part of the wave in the vicinity of hole B will be completely disturbed by the interaction of the particle with the nearby microphysical system. As a consequence of this perturbation, the phase relations between the waves coming from A and B, which are necessary for the appearance of Young's fringes on the film, will not be satisfied.

The foregoing considerations could probably be extended to all interference experiments for every kind of particle. We thus arrive at the following conclusion: "In any interference device, it is impossible to obtain simultaneously both interference fringes and to follow the motion of particles in the interference field." But it can by no means be concluded that this path does not exist. Only the following conclusion may be drawn: "In order that an interference device may function, it is necessary that (1) all paths predicted for propagation of the wave should be actually intercepted by it and that, as a result, the corresponding stream lines should be undisturbed possible trajectories for the particle, and (2) the particle in its trajectory in the interference field should not give rise to any observable localisation phenomenon which would reveal its presence." It is quite obvious that if these two conditions, enabling interference effects to be obtained, are satisfied, the trajectory of the particle, even if it actually exists (as we assume), will not be detected.

We can summarize our point of view by saying that the impossibility of detecting the trajectory of a particle in an interference field results from the intimate bond established by the guidance between propagation of the wave and the displacement of the particle, and not by the non-existence of this or any other mysterious complementarity.

7 – Concerning an Article by Max Born

In concluding this chapter, I should like to say a few words about certain ideas, which Max Born put forward on many occasions, but especially in a recent article in the *Journal de Physique* [10].

It can be accepted that even in classical mechanics the initial conditions for the position and velocity which determine the ultimate motion of a body are never known with complete precision. As a consequence of this uncertainty, which affects the initial data at time t_0, the position of a body at time t can only be predicted with some uncertainty. However, it can be shown that apart from certain very special motions (for example, a linear oscillator), the uncertainty which thus exists in the predicted position increases with $t-t_0$ so that it is possible even from this point of view to introduce into classical mechanics a probability of presence of a body, for example of a material point, which extends to a region of space more and more vast as time passes. Thus, a small uncertainty in the initial position of a body produces a large uncertainty in the position of this body at the end of a sufficiently long time. A long time ago, Henri Poincaré emphasized this fact. Whilst studying the motion of the minor planets over the celestial sphere, he showed that if there were some uncertainty in the position and velocity of a minor planet at a time t_0, then at the end of a sufficiently long time there would be an equal probability that the minor planet might be at any point of the zodiac.

Max Born, after having given precise examples with supporting calculations in order to show that this is true, considers this circumstance as proving that the ideas involved in the usual interpretation of wave mechanics have already been encountered in classical mechanics and that, consequently, they should not appear so extra-ordinary.

For my part, I have treated this conclusion of Max Born with considerable caution. Actually, it is always accepted in classical mechanics that a body has, at every instant, a well-defined position, and if a small uncertainty in its initial position and velocity can give rise to a very large uncertainty on its ultimate position, then there is an uncertainty of "prediction by calculation"; observation proves that the body has, none the less, a well-defined position at every instant. Let us take the problem of Poincaré's minor planets: while the mathematician who is immersed in his calculations of celestial mechanics can no longer say at which point of the zodiac the minor planet is to be found, the astronomer in his observatory looking into his telescope, will always find it at a fixed point with a high degree of accuracy in its trajectory. The mathematician, starting from some initial uncertainty of position and velocity, is led to introduce a probability of presence which results from his not knowing the exact position, and this probability of presence appears to him to extend with time because his lack of knowledge of the position increases. But this does not prevent the position from existing at any instant, and the probability which has been introduced and which represents the lack of knowledge of an existing position is of a wholly classical type. If now, with Max Born, we consider the extent of the uncertainties in classical mechanics as being similar to that in quantum mechanics, we are led to a "double solution" theory, that is to say a theory which supports the idea of a permanent localisation of the particle with time and interprets the probabilities in a classical manner.

Now, we know that this is not at all the point of view of the usual interpretation of wave mechanics. According to this viewpoint, the particle is present in a potential manner throughout the entire extent of its wave train (Born himself emphasizes this potential presence at the conclusion of his article in *Journal de Physique*), and the probabilities which it introduces cannot be interpreted in a classical manner and do not satisfy the formulae which are normally accepted in statistics.

It is a very paradoxical state of affairs that examples drawn from classical mechanics should be used as arguments in favour of a theory which rejects all classical concepts. Born himself was very conscious of this fact, and pointed out repeatedly that the analogies which he used should not be allowed to obscure the fundamental difference which exists between classical and quantum mechanics.

For my part, I think we can come to the following conclusion. Either the analogies between classical and quantum mechanics, on which Max Born insists, are real and they must lead us to an interpretation of the type of the Theory of the Double Solution, or else since the two branches of mechanics are essentially different in their concepts, these analogies are not real and it is not very clear how they serve to elucidate the problem.

Finally, and this is somewhat curious, the analogies urged by Born could be considered as more favourable to a theory of the "double solution" type than to the present-day interpretation.

CHAPTER VI

Wave Mechanics of Systems of Particles and the Theory of the Double Solution

1 – Statement of the Problem

Since the work of Schrödinger in 1926, it has been accepted in wave mechanics that the motion of a system of interacting particles can be represented in the non-relativistic approximation, by the propagation of a wave in configuration space of the system, the space being constituted by the aggregate of the coordinates of the N particles forming the system. The quantity $|\Psi|^2$, which is the square of the amplitude of the wave function Ψ in configuration space, multiplied by an element of volume $d\tau = dx_1 \ldots dz_N$ of this space gives the probability of the *simultaneous* presence of particle 1 in the element of volume $d\tau_1 = dx_1 dy_1 dz_1$ of the physical space, particle 2 in the element of volume $d\tau_2 = dx_2 dy_2 dz_2$ of the physical space, and so on. The success of this method of calculation in very different fields, and particularly in innumerable applications of quantum chemistry, leave no doubt as to its accuracy within its range of validity.

Nevertheless, despite these successes, the wave mechanics of systems of particles in configuration space is of a somewhat paradoxical nature. Firstly, even though it is included within the general framework of a theory which disowns permanent particle localisation in physical space, it deals with an abstract space formed by the coordinates of the various particles in the system. Now, how can an intelligible picture be obtained with the coordinates of a particle which is not localised in physical space? On the other hand it is difficult to accept that the motion of a system of particles can only be described in the abstract hypothetical framework of configuration space and cannot be represented in a physical space of three dimensions. These difficulties do not arise in classical mechanics where the configuration space if defined in terms of the coordinates of material points which are supposed to be completely localised in physical space. The configuration space is frequently used as a convenient means of representing the development of a system of interacting particles, but there is no doubt that the motion of these particles is executed in physical space. On the contrary, in the wave mechanics of systems of particles, the use of the representation in configuration space is obligatory and has no counterpart in physical space. This is indeed strange, and we shall return later to the proof of this fact which was given by Darwin in an early paper.

We must now examine the problem from the point of view of the Theory of the Double Solution which must obviously agree with the representation of systems in physical space, but which must also explain the success of calculations by Schrödinger's method in configuration space.

In the Theory of the Double Solution, every particle in a system constitutes a very small singular region which is incorporated in a wave propagating in physical space. Apart from Bohm–Vigier perturbations, this region executes a motion along one of the stream lines of the wave. The propagation of each individual wave is continuously affected by interactions with the other particles in the system, which is expressed in his individual wave equation by the presence of interaction terms. The aggregate of the trajectories of the particles of the system thus "correlated" can clearly be represented by a single trajectory in configuration space. We are thus led to the fact that the latter trajectory may be identified with one of the stream lines of the complex Schrödinger wave function Ψ in configuration space. However, a consideration of the Theory of the Double Solution will show that the representation of the motion of the system by the Schrödinger Ψ-wave in configuration space is, of course, an "incomplete" representation: in fact, if the stream lines of a Ψ-wave represent exactly all the possible motions of particles belonging to a system in physical space, the Ψ-wave does not represent, in their entire range, the various wave propagations which are associated with these particles in physical space.

The Theory of the Double Solution thus faces the difficult task of analysing precisely the relationships which must exist between particle motion and the propagation of the associated waves which are correlated in physical space on the one hand, and on the other hand, the representation of the motions correlated by the Ψ-wave in configuration space. In a paper published in 1927, I made a first attempt to resolve this difficulty, but on returning to these ideas in 1952–1953, it became apparent that all these attempts were inadequate. However, the problem is one of great importance for the Theory of the Double Solution since, in order to be able to accept the description which it puts forward for the motion of a system (N wave propagations correlated in physical space, each including a singular region), it is essential that this description should explain the success of Schrödinger's wave mechanics in configuration space. One of the main objections which have been made recently, particularly by Fock, to the attempts which have been made to reestablish particle localisation in physical space, is based on the so-called necessity of using configuration space to represent the motion of a system of particles in wave mechanics. A very thorough study of the problem is therefore necessary. Andrade e Silva and I have, for some years now, been pursuing this study, and I shall now summarize the results which we have obtained so far.

2 – *Present State of the Problem. Andrade e Silva's Thesis* *

The principal results may be simply explained by considering the case of a system consisting of two particles only. Generalisation to the case of more than two particles can usually be made without much difficulty.

Firstly, we must make a very important statement. In the Theory of the Double Solution, when we consider the motion of a two-particles system, there are two correlated trajectories T_1 and T_2 in physical space. These two trajectories are the stream lines of the two correlated waves O_1 and O_2. If we now consider some other motion of the system in which the two particles describe two other trajectories T'_1 and T'_2, the latter become the stream lines of the waves O'_1 and O'_2 *which are different from* O_1 and O_2. This results from the fact that the propagation of each individual wave is influenced by the motion of the particle associated with the other wave. There is an essential difference between the case of a particle in motion in a given external field and that of two interacting particles. In the first case, all the stream lines in physical space are the possible trajectories of the particle, whereas in the second case, with each pair of correlated waves in physical space, O_1 and O_2, there is a single pair of stream lines which are the possible trajectories.

We shall now attempt to establish a satisfactory relationship between our picture of a system in physical space and the representation provided

---

* See bibliography [11].

by wave mechanics in configuration space. If we denote by $\mathbf{r}$, $\mathbf{r}_1$ and $\mathbf{r}_2$ the position vectors in physical space of a running point and of the two particles respectively, then the waves associated with the particles can be represented by the formula

$$\Psi_1(\mathbf{r}, \mathbf{r}_2, t) = a_1(\mathbf{r}, \mathbf{r}_2, t)\, e^{\frac{2\pi i}{h} \varphi_1(\mathbf{r}, \mathbf{r}_2, t)},$$
$$\Psi_2(\mathbf{r}, \mathbf{r}_1, t) = a_2(\mathbf{r}, \mathbf{r}_1, t)\, e^{\frac{2\pi i}{h} \varphi_2(\mathbf{r}, \mathbf{r}_1, t)}. \tag{1}$$

The application of the guidance formula to the two waves gives

$$\mathbf{v}_1 = -\frac{1}{m_1}[\operatorname{grad} \varphi_1]_{\mathbf{r}=\mathbf{r}_1}; \qquad \mathbf{v}_2 = -\frac{1}{m_2}[\operatorname{grad} \varphi_2]_{\mathbf{r}=\mathbf{r}_2}. \tag{2}$$

The pair of correlated trajectories T_1 and T_2 of the two particles is represented in six-dimensional configuration space corresponding to the system by a single trajectory T of a hypothetical point, which must be one of the stream lines of the Schrödinger wave function:

$$\Psi(\mathbf{r}_1, \mathbf{r}_2, t) = a(\mathbf{r}_1, \mathbf{r}_2, t)\, e^{\frac{2\pi i}{h} \varphi(\mathbf{r}_1, \mathbf{r}_2, t)}. \tag{3}$$

It can easily be shown that the guidance formula can be used in the form

$$\mathbf{v}_i = -\frac{1}{m_i}[\operatorname{grad}_i \varphi] \qquad (i = 1, 2), \tag{4}$$

where $\operatorname{grad} \varphi$ is a vector in configuration space with the six components $\partial \varphi/\partial x_1, \ldots, \partial \varphi/\partial z_2$.

Comparison of (2) and (4) suggests that the following relationships can be established between the phases φ_1, φ_2 and φ:

$$\operatorname{grad}_1 \varphi = [\operatorname{grad} \varphi_1]_{\mathbf{r}=\mathbf{r}_1}; \qquad \operatorname{grad}_2 \varphi = [\operatorname{grad} \varphi_2]_{\mathbf{r}=\mathbf{r}_2}. \tag{5}$$

Thus, by knowing the values of φ_1 and φ_2 for the correlated trajectories T_1 and T_2 in physical space, it is possible to calculate the phase φ for the trajectory T in configuration space.

The comparison between the individual equations of continuity in physical space and the equation of continuity in configuration space shows that if $a_1(\mathbf{r}, \mathbf{r}_2, t)$ and $a_2(\mathbf{r}, \mathbf{r}_1, t)$ are the amplitudes of the individual waves in physical space, then the amplitude of the Ψ-wave in configuration space is

$$a(\mathbf{r}_1, \mathbf{r}_2, t) = a_1(\mathbf{r}_1, \mathbf{r}_2, t)\, a_2(\mathbf{r}_2, \mathbf{r}_1, t). \tag{6}$$

A detailed analytical proof of this formula can be given, but it can be established more simply in the following way. Let $d\tau_1$ and $d\tau_2$ be two elements of volume, and suppose that they follow individual trajectories T_1 and T_2 in accordance with the guidance formula. The shape and size of these two elements of volume will vary with time, but the individual continuity relationships, which in physical space can be written in the form

$$\frac{D}{Dt}(a_1^2 d\tau_1) = 0 \quad \text{and} \quad \frac{D}{Dt}(a_2^2 d\tau_2) = 0,$$

move in such a way that the quantities $a_1^2 d\tau_1$ and $a_2^2 d\tau_2$ are conserved along T_1 and T_2. To the elements of volume $d\tau_1$ and $d\tau_2$, there corresponds in configuration space a six-dimensional element of volume $d\tau = d\tau_1 d\tau_2$, which according to the guidance formula (4), describes the trajectory T. Now, the equation of continuity corresponding to the Ψ-wave in configuration space, which may be written as $(D/Dt)(a^2 d\tau) = 0$, shows that the quantity $a^2 d\tau$ is conserved along the trajectory T. Thus, it is immediately seen that by adopting the definition given by (6), the conservation of $a^2 d\tau$ along T is a consequence of the conservation of $a_1^2 d\tau$ and $a_2^2 d\tau_2$ along T_1 and T_2.

Finally, if the individual wave functions Ψ_1 and Ψ_2 for the correlated trajectories T_1 and T_2 are known, then equations (5) and (6) enable us to calculate the function Ψ along the corresponding trajectory T in configuration space. The transition from physical space to configuration space can now be represented as follows. If we start with the two waves O_1 and O_2 which in physical space are associated with the correlated trajectories T_1 and T_2, and if we vary the initial conditions in a continuous manner, then we obtain an infinite number of correlated waves O_1 and O_2 which correspond to an infinite number of the correlated trajectories T_1 and T_2. Transition from physical space to configuration space will entail the "removal" from each pair of correlated waves the corresponding trajectories with the values a_1, a_2, φ_1 and φ_2, which they contain (as well as possibly the values of certain derivatives of these functions), and the construction of the trajectory stream lines in the configuration space (by the combination of these individual trajectories), with the Ψ given by equations (5) and (6). It can be seen that this gives an incomplete picture of what happens in physical space, since for each pair of correlated propagations O_1 and O_2 we only preserve that which is associated with the correlated trajectories T_1 and T_2, in configuration space, whilst discarding the remainder of O_1 and O_2.

The Ψ-wave thus obtained from the individual waves Ψ_1 and Ψ_2 with the aid of (5) and (6) must obey the Schrödinger wave equation in configuration space. This means that the individual wave equations of the two particles in physical space must contain "quantum potential of interaction" terms besides the classical and quantized individual potential terms. This is natural enough, and since every particle experiences a reaction, with its own wave expressed by the individual quantized potential, it may also experience reactions due to the waves associated with other particles in the system. These reactions are expressed by the quantum potentials of interaction. The form of these quantum potentials of interaction can be found easily for the case of two particles. For the case of more than two particles the problem is more difficult, but Andrade e Silva has succeeded in solving it, by determining the form of the quantum potentials of interaction in the general case in an unambiguous manner (cf. note in *Comptes Rendus* and Doctoral Thesis [11]).

However, we now encounter the same difficulty as in the case of a single particle. The fact that the quantity $a^2 d\tau$ remains constant along a stream line in configuration space does not enable us to state that it is a measure of the probability of the presence of the representative point of the system within the element of space $d\tau$. Here again we naturally try to introduce the hypothesis of random perturbations of the Bohm–Vigier type, which are such that the representative point continuously moves from one undisturbed trajectory to another in configuration space, and traverses very rapidly all sections of the undisturbed trajectories. A new complication now arises in comparison with the case of a single particle. The Bohm–Vigier perturbations must, clearly, be regarded as acting on the particles in physical space, and they must be introduced into the individual wave equations in the form of perturbation potentials. In the case of a single particle in a given field, all the stream lines for a given wave may be considered as non-perturbed possible trajectories and the argument can be applied to the rapid random transitions from one trajectory to another. In the case of a system in interacting particles, every perturbation of the motion of one of the particles has an immediate effect on the motion of the other, and the pairs of correlated trajectories are the propagation stream lines of *different* waves. This circumstance shows that the arguments of Bohm and Vigier should be thoroughly re-examined again and this has recently been done by Andrade e Silva, especially in his Doctoral Thesis. He has reached a satisfactory conclusion concerning the statistical significance of the quantity $a^2 = |\Psi|^2$ in configuration space.

These considerations, although they still require to be checked and completed, seem to reconcile the statistical predictions of wave mechanics in configuration space with particle localisation in physical space. This removes one of the more important objections against the re-introduction of particle localisation.

3 – *On an early Paper by Darwin*

The interesting paper of Darwin [12] which we shall now discuss, was intended to show that the use of configuration space cannot be avoided in collision theory, since the state of affairs cannot be represented in physical space.

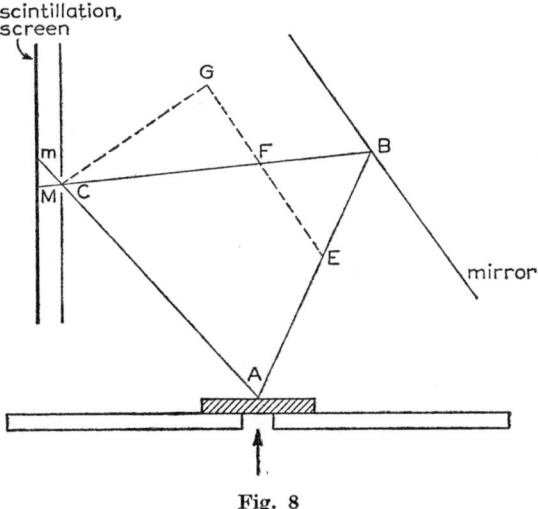

Fig. 8

Darwin considered a plain screen pierced with an opening and covered by a very thin film of homogeneous material whose atoms have mass m. A stream of particles of mass M and velocity V is incident normally as indicated by the arrow in Fig. 8.

We shall suppose that the atoms of mass m and M are able to exert forces on each other within a very short radius of action, i.e. that they collide.

First of all, we shall describe the observable phenomena on the basis of classical physics. At the instant of collision between an incident M atom and an m atom which was at rest in the film, both particles are projected upwards in such a way that energy and momentum are conserved. As a result, M may be projected in the direction AB and m in the direction AC. The velocities of both particles are precisely defined. After the collision there are thus two particles with inter-related motions, which Darwin called a "coherent pair" of particles. Let us suppose that, by means of a mirror, the M atoms coming from A along AB can be reflected to the point C on the trajectory of the m atom which came from A along AC.

We now arrange a second screen at C, also pierced with a hole, and behind this screen we instal a scintillation screen which records the arrival of the particles. In general, we shall observe on this screen the arrival of M-atoms along AB at the point M, and also the arrival of m-atoms along AC at the point m. In a very special case, however, in which after the collision the M-atoms possess a velocity considerably greater than the velocity of m-atoms, the time taken by the former to traverse the trajectory ABC may turn out to be the same as the time taken by the m-atoms to go directly from A to C. The resulting collision at C of the M- and m-atoms is such that the scintillation screen will be able to receive the atoms at points other than M and m. Thus, for a certain position of the mirror and of the point B, a quite different phenomenon can be produced from that observed at other positions. This is what the classical picture portrays.

Darwin then examines how the same phenomenon could be interpreted by introducing waves associated with the particles in physical space, but, (and here is the essential point) he accepts that these waves are continuous and of the usual type of Ψ-waves. Consequently, they will not enable us to define the location and distance of the two particles. He thus associates with each M- and m-atom a spherical homogeneous wave propagating from A. The fact that a particular phenomenon is produced for a certain position of the mirror thus appears to him to be incapable of interpretation by interference produced at C between the M-wave coming from A by reflection at B and the m-wave coming directly from A. Actually, this interference could only be produced if there were a certain well-defined phase relation between the waves which cross at C. However, we can place mirrors at E and F in such a way that the wave associated with the M-atom arrives at C after two reflections at E and F and thus has the same phase as before. For this, the path difference EBF—EF must be equal to an integral number of wavelengths of the M-wave. The conditions for interference at C will then be the same, but it is certain that collisions at C would no longer occur because the M-atoms arrive at C before the m-atoms. On the contrary, if one mirror is placed at E and the other at G so that the path AEGC is equal to the path ABC, collisions will always occur at C, whilst the conditions for interference at C will be completely modified, since the waves will arrive there through a different angle than that in the original experiment. Finally, from the point of view of intensities there would also be a difficulty, because it is obvious from the particle point of view that the probability of a collision at C must be inversely proportional to $(\overline{AC})^2$, whilst interference at C will give an intensity proportional to the product of the intensities of the interfering waves or to the fourth power of $1/\overline{AC}$.

Darwin concluded from this that the phenomena cannot be represented by associating with each particle a wave propagating in three-dimensional space. On the contrary, the phenomena *can* be represented by associating a system of two particles M and m with a single wave Ψ in six-dimensional configuration space. Darwin carried out the complete calculation and found in effect that it is possible to represent the fact of the second collision of the atoms at C only for a certain position of the mirror by using the propagation of the wave Ψ in configuration space. Darwin claimed that this example illustrates the necessity of using the configuration space in the wave mechanics of systems, and in particular, in collision problems.

Darwin's demonstration obviously proves the impossibility of representing collision phenomena, which are studied by means of the propagation of two homogeneous waves without singularities, in three-dimensional physical space. This is not particularly surprising since Darwin accepted *a priori* that the results must conform to the concept of well-localised particles in space and that the homogeneous waves contain nothing that will enable this localisation to be defined. Darwin's reasoning thus shows only that the wave associated with a particle in three-dimensional physical space cannot be a homogeneous wave and that it must contain an inhomogeneity which enables its localisation to be defined. This is entirely in accordance with the concepts of the Theory of the Double Solution.

Moreover, it can easily be seen why the collision problem can be dealt with correctly by the configuration space method. It is because, as Darwin pointed out (pp. 388–389 of his paper), it introduces a potential term $V(\mathbf{r}_1-\mathbf{r}_2)$ into the wave equation in configuration space, which represents a very short range interaction (when the position vectors $\mathbf{r}_1$ and $\mathbf{r}_2$ of the two particles are very nearly the same). Clearly, the introduction of a similar term is impossible when we consider the propagation of two *homogeneous* waves in physical space. If the configuration space method is successful in dealing correctly with the problem, it is because it establishes a particle localisation in physical space *subreptitiously and without admitting it*. In point of fact, the variables $\mathbf{r}_1$ and $\mathbf{r}_2$ represent nothing but the positions of the particles in physical space. In other words, the use of configuration space implies particle localisation in physical space. In order to obtain a picture of the collision phenomenon in physical space, the concept of homogeneous waves associated with particles cannot be used: it does not enable the distance between the two particles to be defined. Only the phase and amplitude differences could be used to explain the phenomenon, and Darwin has clearly shown that this would not be sufficient. But this is not so if each particle in physical space is associated with a wave which contains a very localised singularity or singular region, which enables the position of the particle to be defined within its wave, and consequently, to introduce into the individual wave equations of a two-particle system a term $V(\mathbf{r}_1-\mathbf{r}_2)$ expressing a very short range interaction.

Finally, we arrive at the quite unexpected conclusion, that Darwin's article, far from showing the necessity of dealing with collision problems in configuration space, in fact reinforces the argument that the wave associated with a particle in physical space must contain a very small singular region, and this leads us back to the u-wave of the Theory of the Double Solution.

CHAPTER VII

Remarks on Systems of Identical Particles*

1 – General Considerations

The theory of systems of particles which we have investigated so far, has been concerned with ensembles of distinguishable particles with zero spin. In addition to being of interest in itself, this analysis suggests that a description in physical space of any type of particles *with spin* seems to be possible—as the Theory of the Double Solution suggests.

Any such generalisation must involve the solution of at least two problems. Firstly, it must explain why the equivalence of particles in a system involves the symmetric or antisymmetric property of the wave function Ψ in configuration space, and secondly, it must elucidate the origin of the mysterious exclusion principle. At present, research into the properties of systems of particles with spin has not advanced far enough for us to make more than a few comments of a rather general nature in connection with Pauli's principle. We shall therefore confine our attention to systems of identical spinless particles in attempting to sort out the physical meaning which, according to the Theory of the Double Solution, seems to be connected with the symmetrization of the wave function Ψ.

The casual interpretation of wave mechanics by the Theory of the Double Solution is based on the completely deterministic dynamics of regular waves containing singular regions. A kind of stochastic principle is introduced (the postulate of random fluctuations) which is supposed to relate the weak random interactions of the microsystem with its environment. This allows the deduction of a statistical theory which is valid both for single particles and for systems of particles, and provides us with the usual probabilistic interpretation of the square of the modulus of the wave function Ψ. However, the identical nature of the particles

---

* This chapter was written by J. Andrade e Silva.

in the system requires a change in the form of the wave function in the corresponding configuration space. Thus, when bosons are concerned, the wave function must be symmetric in the coordinates of the various particles. Since the Theory of the Double Solution makes the form of the function Ψ dependent on the basic description in physical space, it follows that this description must also undergo a corresponding change. We are thus first confronted with the problem of knowing whether, in relation to the formalism developed for systems of different particles, the necessary change still has an effect at the level of the deterministic dynamics of the waves, or whether on the contrary it only expresses the qualitatively different effects of the random fluctuations when they act on systems of identical particles.

This first problem is directly related to the problem of the significance of the symmetry of the wave function Ψ. We know that the usual explanation amounts to saying that since the particles are physically identical and have in general no precise location, it is not possible to determine experimentally which particle has been detected at a certain point in space at any given instant. Since we demand of theory the exclusion of any prediction which cannot be verified experimentally, it can be concluded that the formalism must be confined to giving the probability that a *single* particle may be found at a certain instant in a certain element of volume. We must thus restrict ourselves to the use of forms of the function Ψ which furnish predictions when the initial numbering has been erased, i.e. to symmetric or antisymmetric forms.

It goes without saying that the causal interpretation cannot accept this "justification" for the invariance of $|\Psi|^2$ with respect to permutation of the particle coordinates. New physical properties which are specific for systems of identical particles must replace a methodological requirement. Nevertheless, the interpretation of symmetrisation of the function Ψ in terms of the erasure of the initial numbering of the system (a system of bosons in this particular case) clearly expresses an objective physical reality. On the other hand, the Theory of the Double Solution establishes, in principle, a localisation which only evades us in practice because of the effects of small random fluctuations. As a result, whilst the existence of these fluctuations is not taken into consideration, the localisation and hence the numbering of the particles maintain all their significance. In other words, erasure of the initial conditions and the correlative symmetrisation of the function Ψ only take place as a consequence of the effect of small random perturbations of a system of identical particles. Thus, we must take for a system of identical particles the same equations of motion as for a system of different particles, and we must consider the symmetrisation of the function Ψ as representing a physical effect which results solely from the sameness of the physical nature of the particles of a system subjected to perturbations.

In order to define the new physical property corresponding to symmetrisation, let us consider again the picture which the Theory of the Double Solution provides us for a system of particles. We now have an interaction (classical or quantum) between a group of waves v each one of which carries a singularity u_0, such that it is always in phase with its wave. The configuration space in which the normal function Ψ evolves is thus based on the coordinates of the singularities. In this model, therefore, the symmetrisation rule for the function Ψ corresponds necessarily to the possibility of exchanging the positions of the singularities. More precisely, and in full agreement with what has just been stated, we must accept that a singularity may pass from a wave v to another wave of the same physical nature because of the action of random perturbations.

We shall show later that it is this additional possibility which characterises, essentially, systems of identical particles, since it enables the symmetric nature of the state function to be predicted when spinless particles are involved. However, we must first discuss in detail its physical content and in this connection introduce a number of general comments.

The idea that a singularity may pass from a wave v to another wave of the same physical nature, evidently constitutes a new hypothesis in the Theory of the Double Solution. It corresponds to defining the nature of the bonds existing between the regular waves v and the singularity waves u_0, and this suggests that the two types of waves are linked essentially by their identity. This agrees all the better with the basic concepts of the theory which indicates that identical singularities correspond to identical particles, but it is not clear why a singularity should be associated with a particular v-wave in preference to another. On the other hand, it seems necessary, in our model, that the transfer of the singularity from one v-wave to another should only take place if there is a total or partial superposition of these two waves at the instant of transition. Now, experiment shows precisely that the symmetrisation rules for the function Ψ must only be applied if there is a preliminary overlap between the wave trains corresponding to the particles under consideration. Thus, a simple and natural explanation is obtained of an experimental fact, but the purely probabilistic interpretation of this fact is somewhat subjective in nature.

2 – Introduction of the Concept of Transitory States

Nevertheless, this new hypothesis on the behaviour of the singularities seems to raise important difficulties. For example, it poses the problem of knowing why the transition of a singularity from one wave to another of the same type should result from the action of random perturbations whilst at the same time not playing any part in the underlying dynamics. Similarly, we must be able to understand the reasons for observing, once the final state is attained, only one singularity on each regular wave.

These problems are associated with the description of the physical process which is responsible for the transition of the system from the equilibrium state which precedes the interference of the v-waves to a new state of equilibrium which follows this interaction (and which corresponds to a Ψ-function which now becomes symmetrical). We cannot give a satisfactory answer to this problem and this is a consequence, we believe, of the fact that they are, to a large extent, outside the present-day framework of wave mechanics. Thus, in view of the importance which could be attached to this conclusion, we shall continue further with the analysis of the considerations which seem to support it.

De Broglie was the first to emphasize, with Einstein, the necessity of considering the present formulation of wave mechanics as fundamentally incomplete; he has put forward over a long period his convincing arguments, based on the Theory of the Double Solution, in favour of introducing non-linear terms into the wave equations. In his opinion, non-linear effects could, in certain instances, play an essential role, as would certainly be the case for the relationships between regular waves and singular regions. A close connection with the problem that concerns us here can already be seen.

However, more recent papers * treat the problem of the incomplete nature of the usual formalism from another point of view. In view of the inability of a purely linear theory to furnish a relative determination

---

* J. Andrade e Silva, F. Fer, P. Leruste and G. Lochak, *Compt. Rend.* (*Paris*), 251 (1960) 2305, 2482, 2662; *Cahiers Phys.*, 15 (1961) 210 and 16 (1962) 1.

of amplitudes (essential, as we know, in wave mechanics) and the necessity for all quantum theories to introduce *a posteriori* non-linear postulates, it seems advisable to start with non-linear equations. We could thus obtain, as suggested by the examples quoted above, a more general theoretical scheme capable of describing transitory states as well as "stationary" states and the latter would only appear as the necessary outcomes of transition processes.

In particular, this would amount to assigning a fundamental significance to transition states, the description of which cannot be confidently obtained from linear and Hamiltonian dynamics.

Until now, any research work undertaken in this domain (and we do not propose to summarise it here) has been concerned with quantized systems. The "stationary" states of the system (here the word "stationary" has, as before, a physical meaning) were thus taken to correspond to the experimentally observed quantized states. But one can try to go further and apply the same kind of ideas to the propagation problems of wave mechanics. Thus, equilibrium statistics would correspond to "stable" states (i.e. states which are not of a very short transitory nature), whilst transition processes would be related to the events necessary for attaining these equilibrium states.

Without ignoring the basic differences between these two types of problem, it is evident that this generalisation has important advantages quite apart from the interests of coherence. In anticipation of a detailed treatment, for which there is no room here, we shall confine ourselves for now to two simple statements. Firstly, research into quantized systems has led to the assignment of certain ergodic properties to the corresponding stationary states. These final states, generally finite in number, would thus correspond unequivocally to continuous sets of initial conditions. Should one not establish a relationship between this ergodism and that which clearly represents the attainment of the equilibrium distribution for $|\Psi|^2$? Finally, if Hamiltonian dynamics is effectively incapable of describing the transition from one quantized state to another, and if similarly (we shall return to this later), it cannot describe the physical process which leads to the $|\Psi|^2$ distribution, should one not regard this merely as a coincidence? On the contrary, is it not necessary to use non-linear and non-Hamiltonian mechanics to describe all these ultra-rapid processes, whether they are quantized transitions or transitions from arbitrary initial states leading to the $|\Psi|^2$ probability distribution?

Now, if these analyses are justified, it follows that the problems which we encountered prior to this long digression are really outside the present theoretical framework, since they are concerned with the description of a transition process. Of course, it could be objected that this conclusion only results from a theoretical scheme which is, on the whole, quite hypothetical. This is why we shall reconsider this problem from a much more direct point of view in order to show quickly the importance which can be attributed to the description of these transition states and to the incomplete nature of the present-day theory.

Leaving aside the usual interpretation of wave mechanics where the distribution of the probabilities $|\Psi|^2$ is postulated, and quantized transitions are considered as "indescribable", we shall commence by recalling that the linear form of the Theory of the Double Solution is no longer capable of describing the physical process which leads to the attainment of the $|\Psi|^2$ distribution of probabilities. Whether this be in the case of a single particle or in the case of a system of particles of different types, it succeeds well in deducing the existence of this probability distribution starting from one physical principle, i.e. the concept of random perturbations. However, this deduction, which depends nowadays on the ergodic theorem of Markov chain theory, gives us no information whatsoever about the evolution of the system from its initial state (which can be, *a priori*, anything whatsoever) up to the final equilibrium state. In other words, the description of the transition process itself completely escapes us and we are left only with the identification of the state in which the process must necessarily terminate.

There is no reason whatsoever for supposing that things would happen otherwise if a system of identical particles were concerned, nor that in this particular case it would be possible to give a description of the transition states. On the contrary, now we are told that the system shows more remarkable properties, we can hope to highlight the difficulties raised by the hypothesis that such a description of a system of identical particles is actually "complete".

In fact, every possible interpretation of the formalism of wave mechanics will take into account the fact that symmetrisation of the Ψ-function of the system, in consequence of the interference of the individual wave trains, conveys in some manner or other the possibility of every particle being present in a region of space where it was certainly not found previously. Now, the essential feature is that symmetrisation corresponds to the realisation of a certain physical process, whether it be supposed describable or not. It seems obvious to us, however, that such a physical process (just as any other) cannot take place in strictly zero time. Thus the first indication that theories based on linear equations are incomplete lies in their failure to predict the duration of this phenomenon.

Suppose now that $\Psi(1, 2)$ is the state function which represents a system of two identical particles prior to the interference of the individual wave trains. We know that, because of the symmetry of the Hamiltonian of the system, $\Psi(2,1)$ is also a solution of the Schrödinger wave equation in configuration space and we can write the solution in the form of the linear combination $A\Psi(1, 2) + B\Psi(2, 1)$, where A and B are arbitrary constants which are related by the normalisation rule for the state function. Now, of all the elements in this continuous family of solutions, the present-day theory only retains two possibilities which correspond respectively to taking $B = 0$ (and hence, after normalisation, $A = 1$), and $B = A$ (their common value thus being $1/\sqrt{2}$). But this transition, which is necessarily discontinuous, from the first type of solution to the second, which expresses symmetrisation of the state function, seems to be incompatible with the concept of a physical process having a finite duration. It is not clear why other solutions corresponding to values of A lying between 1 and $1/\sqrt{2}$ and the corresponding values of B lying between 0 and $1/\sqrt{2}$ should not, under certain circumstances, be taken into consideration; they would thus express the possibility of a "partial" exchange of particles (the possibility of a total exchange corresponding to the solution $A = B = 1/\sqrt{2}$ and an impossibility of exchange to $B = 0$ and $A = 1$). Moreover, it could very well be that the properties of these intermediate states would not be given by this linear mechanics, which is equivalent to saying that they would not be described by the functions $A\Psi(1, 2) + B\Psi(2, 1)$ where $1 > A > 1/\sqrt{2} > B > 0$. We suggest that this would then show that we should have to deal with transitory states, which could only be described by starting from a more general mechanics.

These ideas concerning systems of identical particles are not perhaps quite so speculative as they may appear to be. We wonder whether it would not be possible to determine experimentally the distribution of probabilities corresponding to wave trains which have previously "interfered" and which, nevertheless, are not reconcilable with the normal symmetrised wave function. This is equivalent to assuming that the present success of the strict symmetrisation rule for Ψ is only due to the exceedingly rapid nature of the corresponding transition process, and that under certain special conditions this process could, in some way or other, be interrupted before the end. For example, it might be possible to devise an experimental arrangement whereby two trains of monokinetic waves of quite large dimensions and both composed of particles of the same physical nature could be assigned initial conditions chosen in such a manner that later on they become superimposed very slightly for an extremely short period of time. It would thus be possible to measure on screens the distribution of the consecutive probabilities of the presence of particles in order to establish whether they were in accordance with the predictions based on the state function which is symmetric or asymmetric according to the spin. It does not seem unreasonable to suppose that an experiment of this type would be accurate enough to reveal a new physical fact; in any case it would contribute interesting information about the range of validity of the present theoretical scheme.

3 – *Demonstration of the Symmetry of Wave Functions for Bosons*

In conclusion, we shall confine our attention to an outline of the reasoning which enables us to deduce, in terms of the Theory of the Double Solution, the symmetry of the Ψ-function for a system of identical, spinless particles. To do this, we must first recall a number of general points.

The paper which Bohm and Vigier have devoted to the problem of identification of the probability of presence of a particle with the classical $|\Psi|^2$ distribution in the case of a single particle is based on the physical assumption of the existence of small perturbations of a random nature, which act on the system. Now, the assumptions which must be made concerning the effects of these perturbations (the behaviour of which has been considered by Bohm and Vigier in a hydrodynamic model) amount to comparing the random motion of the singularity on the regular wave with a Markov chain of a type which is well-known in the theory of probability. We know also the "representative space" corresponding to this chain, which is composed at any instant of the spatial region occupied by the regular wave. Despite the absence of precise data on the probability matrix for transitions between the volume elements of the regular wave, nevertheless we know that this matrix exists and this is sufficient for the application of the ergodic theorem in the Markov

Theory. This theorem tells us that for chains of this quite general type (irreducible and non-periodic chains) only one stationary state can exist and that the probability distribution would, in the course of time, tend towards the stationary state, whatever might be the initial distribution. Thus, it was only necessary for Bohm and Vigier to show that the $|\Psi|^2$ probability distribution was effectively an equilibrium distribution (in the Markov sense) in order that the required result might be obtained within wide limits.

There is no need here to recapitulate the reasons which made it necessary to generalise this argument so that it could be applied to systems of particles of different types *. Instead, we would emphasize the fact that this generalisation does not affect the essential points of the Bohm–Vigier demonstration. Here too, a random motion of the Markov type was introduced for each singularity, and the representative spaces associated with the regular waves were defined. It was still Markov's ergodic theorem which assured us of the tendency of the system to resume an equilibrium state which could be identified with the $|\Psi|^2$ function in configuration space.

We now propose to show how, by taking into consideration the possibility of the transition of singularities from one v-wave to another, an identical *mutatis mutandis* reasoning can still predict the symmetry of the Ψ-waves.

Let us suppose that we have a system of two identical, spinless particles, represented by two regular waves on which are moving two singularities. For simplicity, and because this does not involve any loss of generality, we shall suppose that initially the two v-waves do not interfere. We shall number the regular waves arbitrarily (though distinctly) as follows: the singularity which is initially on wave v_1 we shall designate by 1, and the singularity located on wave v_2 by 2. In accordance with our preceding remarks, we must retain exactly the same random perturbations as in the case of systems of different particles until the interference of the v-waves has taken place. We shall now have a Markov-type evolution for each of the singularities with exactly the same representative space (designated E_1 and E_2) as before for each of the Markov chains. From the instant of interference, however, the evolution of each singularity, whilst remaining of the same type, requires a larger representative space; this

---

* See Andrade e Silva's Thesis, Bibliography [11].

enlargement must express the possibility of each singularity transferring to the other regular wave. It can easily be shown that the two particles now occupy identical representative spaces of the form E_1+E_2. Moreover, we could have introduced formally these extended representative spaces from the beginning. In order to take into account the impossibility of the transference of a singularity from one regular wave to another prior to interference, it is sufficient to assume explicitly that until interference takes place, all the matrix elements of the transition probabilities which correspond to the intersection of the sub-spaces E_1 and E_2 are zero. Since singularity 1 corresponds to wave v_1 and singularity 2 to wave v_2, and similarly in the extended representative space, each singularity will be forced to remain on the same wave until interference does take place. From this point of view, the change brought about by the identity of the particles which constitute the system is expressed simply by the fact that the matrix elements corresponding to the intersection of the sub-spaces E_1 and E_2 can become non-vanishing at a certain instant (where there is interference).

Since Markov's ergodic theorem remains essentially valid in this context, we are thus assured that any initial state will always lead to the same final equilibrium state (which we assume to exist). Strictly, we can take a first initial state in the extended representative space where singularity 1 is on wave v_1 (i.e. in sub-space E_1) and singularity 2 is on wave v_2 (i.e. in sub-space E_2), and a second initial state corresponding to the exchange of the positions of the singularities. This second possibility, only apparently contradicts the original numeration, since at some subsequent time the singularities may have a finite probability (because the chain is irreducible) of exchanging their positions, or what amounts to the same thing, we must define the chains on which the ergodic theorem operates in a representative space of the form E_1+E_2, which corresponds precisely to the necessity for taking into consideration the initial conditions of this type. Expressed in the language of physics, this is the same as saying that the possibility of the singularities changing positions after interference of the v-waves now represents the possibility of this exchange commencing at the initial instant.

Let us now compare the two continuous evolutions of the system corresponding to these initial states, which are characterised (relative to each other) by the exchange of the positions of the singularities. Since the two states are physically identical and are only distinguishable by the different notation applied to the singularities, it will be the same during the entire evolution process, up to the final state of stochastic equilibrium. Evidently, the result is that this must correspond to a description which is independent of the initial numbering, i.e. which is invariant with respect to the permutations of the singularity coordinates. Thus, the symmetry of the Ψ-function in configuration space has been demonstrated within the framework of the Theory of the Double Solution.

APPENDIX

Negative Probabilities and the Theory of the Double Solution

In the general theory of wave equations for particles with spin, one is occasionally concerned with densities ρ, which although conforming to the continuity equation, are not positively defined. If we accept, as usual, that the probability of presence $P(\mathbf{r}, t)$ of a particle at a point $\mathbf{r}$ in space at the instant of time t is proportional to $\rho(\mathbf{r}, t)$, then negative probabilities must be introduced and this is hardly acceptable. We propose to demonstrate how the concepts of the Theory of the Double Solution would perhaps enable this difficulty to be avoided.

In accordance with the Theory of the Double Solution, we accept that particles are localised in space at every instant, and that their motion is determined by propagation of the wave to which they are attached. We accept also that they have a velocity $\mathbf{v}$ defined by the guidance formula: this formula can be taken in the form which I normally employ, but it is not necessary to define it more precisely for the following reasons. Knowing the velocity $\mathbf{v}(\mathbf{r}, t)$ we will still accept the existence of a quantity $\rho(\mathbf{r}, t)$ which may be positive or negative and which satisfies the equation

$$\frac{\partial \rho}{\partial t} + \mathrm{div}(\rho \mathbf{v}) = 0. \tag{1}$$

We shall now demonstrate an important result.

Let us consider two points $\mathbf{P}$ and $\mathbf{Q}$ in physical space at a time t, such that $\rho(\mathbf{P}, t) > 0$ and $\rho(\mathbf{Q}, t) < 0$. If we join $\mathbf{P}$ and $\mathbf{Q}$ by any continuous curve, and if ρ is a continuous function of $\mathbf{r}$, then the quantity ρ will vanish at least once on the curve joining $\mathbf{P}$ and $\mathbf{Q}$. Since this result is valid for any pair of points $\mathbf{P}$ and $\mathbf{Q}$ and any t, we can conclude that a surface S exists (which generally becomes distorted in course of time) which constantly separates a region of space D^+, where ρ is positive, from a region of space D^- where ρ is negative. On this surface, the flux

$\rho\mathbf{v}$ is always zero, since ρ is zero. Consequently, no particle can pass through S. Every particle initially contained in D^+ remains in D^+, and every particle initially contained in D^- remains in D^-.

This result can be found in another way. We consider a very small element of physical space $d\tau$ containing a group of material points with velocities $\mathbf{v}$. If ρ denotes the value of $\rho(\mathbf{r}, t)$ inside $d\tau$, then the continuity equation (1) can be written

$$\frac{D}{Dt}(\rho d\tau) = 0. \tag{2}$$

In this form, it can be seen that in course of time the magnitude of the element $d\tau$ containing the material points and also the quantity ρ may vary, but the product $\rho d\tau$ remains constant when the motion of $d\tau$ along a streamline is followed. But the element of volume $d\tau$ always maintains a positive value, and consequently ρ always maintains the same sign along the streamline. Thus, if a particle is initially located in D^+, its trajectory is such that it always remains in D^+ and, if a particle is initially located in D^-, its trajectory is such that it always remains in D^-. We have thus arrived at the result shown above.

We can now make the important comment that the double proof which we have just given depends entirely on the assumption that the particles have a trajectory and that the continuity equation is valid. These proofs now continue to be valid in their entirety if we introduce into the wave equations the perturbation potentials representing the continuous random perturbations which, according to the Bohm–Vigier hypothesis, represent the continuous interaction of the particle with the "subquantum medium" and this extends the generality of the result obtained.

We are now in a position to foresee how the paradox of negative probabilities can be eliminated. The particles present in D^+ and in D^- form two entirely separate groups, and we can define the probabilities of presence $P(\mathbf{r}, t)$ by putting

$$\begin{aligned} P(\mathbf{r}, t) &= C\rho(\mathbf{r}, t) \quad \text{in} \quad D^+, \quad C > 0, \\ P(\mathbf{r}, t) &= C'\rho(\mathbf{r}, t) \quad \text{in} \quad D^-, \quad C' < 0; \end{aligned} \tag{3}$$

where C and C' are normalisation constants. The definitions given in equation (3) enable us to obtain the probabilities of presence which are, and everywhere remain, positive.

It should be noted that this method of overcoming the difficulty of negative probabilities, if it has indeed been demonstrated satisfactorily, has resulted from the assumption that the particles are localised in physical space at every instant of time, and that they have a velocity determined by the guidance formula in such a way that they have well-defined trajectories. Clearly, the usual interpretation of wave mechanics which rejects these assumptions cannot provide a similar method of getting round the difficulty.

Bibliography

[1] L. DE BROGLIE, *Une tentative d'interprétation causale et non linéaire de la Mécanique ondulatoire: la théorie de la double solution*, Gauthier-Villars, Paris, 1956. English translation: *Non-linear wave mechanics, a causal interpretation*, Elsevier, Amsterdam, 1960.

[2] L. DE BROGLIE, *La théorie de la Mesure en Mécanique ondulatoire*, Gauthier-Villars, Paris, 1957.

[3] L. DE BROGLIE, L'interprétation de la Mécanique ondulatoire, *J. Phys. Rad.*, 20 (December 1959) 963.

[4] W. RENNINGER, Measurements without interference from the measurement device, *Z. Physik*, 158 (1960) 417.

[5] D. BOHM and J. P. VIGIER, Model of the causal interpretation of quantum theory in terms of a fluid with irregular fluctuations, *Phys. Rev.*, 96 (1954) 208.

[6] F. FER, J. ANDRADE E SILVA, PH. LERUSTE and G. LOCHAK, *Compt. Rend. (Paris)*, 251 (1960) 2305, 2482, 2662.

[7] G. DARMOIS, *Les équations de la gravitation einsteinienne* (*Mémorial des Sciences mathématiques*), Gauthier-Villars, Paris, 1926.

[8] A. LICHNEROWICZ, *Les théories relativistes de la gravitation*, Masson, Paris, 1955.

[9] L. DE BROGLIE, La statistique des cas purs en Mécanique ondulatoire et l'interférence des probabilités, *Rev. Sci.*, March 1948, p. 259.

[10] MAX BORN, Dans quelle mesure la Mécanique classique peut-elle prévoir les trajectoires?, *J. Phys. Rad.*, 20 (January 1959) 43.

[11] J. ANDRADE E SILVA, *La théorie des systèmes de particules dans l'interprétation causale de la Mécanique ondulatoire* (Ph. D. Thesis). *Ann. Inst. H. Poincaré*, 16, No. 4 (1960) 289–359.

[12] C. G. DARWIN, A collision problem in the wave mechanics, *Proc. Roy. Soc. (London)*, A124 (1929) 375.

[13] L. DE BROGLIE, Sur la Thermodynamique du corpuscule isolé, *Compt. Rend. (Paris)*, 253 (1961) 1078.
Nouvelle présentation de la Thermodynamique de la particule isolée, *Compt. Rend. (Paris)*, 255 (1962) 807.
Quelques conséquences de la Thermodynamique de la particule isolée, *Compt. Rend. (Paris)*, 255 (1962) 1052.
La Thermodynamique de la particule isolée, Gauthier-Villars, Paris (in the press).

PRINTED IN THE NETHERLANDS